# CHINA
## AND GLOBAL CHANGE

## OPPORTUNITIES FOR COLLABORATION

Panel on Global Climate Change Sciences in China

Committee on Scholarly Communication
with the People's Republic of China

Office of International Affairs

National Research Council

NATIONAL ACADEMY PRESS
Washington, D.C. 1992

NOTICE: The program of studies of Chinese science was begun in 1990 to inform the scholarly community about the current state of science inside and outside of China. The program was approved by the Governing Board of the National Research Council, whose members are drawn from the councils of the National Academy of Sciences, the National Academy of Engineering, and the Institute of Medicine. It was supported under Master Agreement Number 8618643 between the National Science Foundation and the National Academy of Sciences and Contract Number INT-8506451 between the National Science Foundation and the Committee on Scholarly Communication with the People's Republic of China (CSCPRC).

Founded in 1966, the CSCPRC represents American scholars in the natural and engineering sciences, the social sciences, and the humanities, including Chinese studies. The Committee is composed of scholars from all of these fields. In addition to administering exchange programs, the CSCPRC advises individuals and institutions on scholarly communication between the United States and China. Administrative offices of the CSCPRC are located in Washington, D.C.

Library of Congress Catalog Card Number 92-62952
International Standard Book Number 0-309-04841-9

Additional copies of this report are available from:

National Academy Press
2101 Constitution Avenue, N.W.
Washington, DC 20418

B-062

Copyright 1992 by the National Academy of Sciences. All rights reserved.

Printed in the United States of America

## PANEL ON GLOBAL CLIMATE CHANGE SCIENCES IN CHINA

JAMES N. GALLOWAY, *Chairman*, University of Virginia
JOSEPH A. BERRY, Carnegie Institution of Washington, Stanford University
ROBERT E. DICKINSON, University of Arizona
C.S. KIANG, Georgia Institute of Technology
SHAW C. LIU, National Oceanic and Atmospheric Administration
WILLIAM A. REINERS, University of Wyoming
DAVID S. SCHIMEL, National Center for Atmospheric Research
NIEN DAK SZE, Atmospheric and Environmental Research Inc.
WEI-CHYUNG WANG, State University of New York, Albany
JOHN W. WINCHESTER, Florida State University

*Staff*

Beryl Leach, Program Officer

## COMMITTEE ON SCHOLARLY COMMUNICATION WITH THE PEOPLE'S REPUBLIC OF CHINA

JAMES D. EBERT, *Chairman*, Marine Biological Laboratory
NICHOLAS R. LARDY, *Vice Chairman*, Jackson School of International Studies, University of Washington
MARY BROWN BULLOCK, The Wilson Center, Smithsonian Institution
ELLIS B. COWLING, North Carolina State University
GERALD P. DINNEEN, National Academy of Engineering
HARRY HARDING, The Brookings Institution
CHARLES E. HESS, University of California, Davis
DWIGHT H. PERKINS, Harvard Institute for International Development, Harvard University
E. PERRY LINK, Princeton University
ROBERT B. OXNAM, The Asia Society
JAMES B. WYNGAARDEN, Institute of Medicine

*Staff*

Robert Geyer, Director

# Preface

Human impacts on the environment transcend political and geographic boundaries. For example, atmospheric emissions from one country can impact ecosystems half-way around the world from it. The scale of these impacts has given rise to the term "global change." However, scientific understanding of these global change processes is very limited, and better understanding and timely responses will depend on international cooperation. To this end, the International Council of Scientific Unions has established the International Geosphere-Biosphere Program and, with the World Meteorological Organization, the World Climate Research Program.

The active involvement and contributions of developing countries to the study of global change are crucial. And China, by virtue of having 22 percent of the world's population and being a significant contributor to carbon dioxide, methane, sulfur dioxide, chlorofluorocarbon, and other greenhouse gas emissions, is one of the most important of these countries. China's emissions are already having a profound impact locally, and regional and global impacts are of growing concern. Moreover, because of China's projected population growth and reliance on high-sulfur coal for 75 percent of its energy needs, a potential exists for substantially greater impacts in the future.

The main purpose of this report is to provide detailed information on Chinese global change research in order to facilitate collaboration that will increase understanding of China's impact on global change and the impact of global change on China. Furthermore, Chinese geography, such as the Loess and Qinghai–Tibet Plateaux and extensive coastline provide fertile territory for investigating processes that regulate the Earth's environment. A wealth of unique historical proxy data offers a unique and significant contribution to studies of past global changes. Most importantly, China has a community of scientists who recognize the importance of global change studies and who are committed to improving our understanding of China and global change.

The study of global change sciences in China was itself a collaborative effort between the panel and our colleagues in China. Without their assistance in gathering information, their patient and extensive explanations of their work and priorities, and their openness and collegiality—the spiritual backbone of this effort—the study would not have been possible. As this report shows, the Chinese global change program is extensive, so I am not able in this space to thank all of the Chinese scientists to whom we are indebted. We sincerely appreciate the efforts and support of Ye Duzheng, chairman of the Chinese National Committee for the IGBP, Lin Hai of the National Natural Science Foundation, Fu Congbin, Shi Yafeng, Chen Panqin, Wang Hui, Cheng Erjin, and An Jianji from the Chinese Academy of Sciences, Ding Yihui of the Chinese Academy of Meteorological Sciences, and Liu Chunzheng of the Ministry of Water Resources. To the institute directors and the scientists at the institutions the panel visited, named and unnamed in the report, we very much appreciate your hospitality and cooperation.

The production of this report required the efforts of many individuals. I am deeply appreciative to the panel members who found time in already overcommitted schedules to travel to China and tackle the job of sorting through the extensive body of information the panel compiled. In particular, my thanks to Dave Schimel and Shaw Liu for organizing Chapter 5. The panel is indebted to the substantial support received from the Committee on Scholarly Communication with the People's Republic of China (CSCPRC), which was absolutely essential for the completion of a successful report. John Olsen, Gao Xing, Keith Clemenger, and Yuan Xiansheng of the National Academy of Sciences/CSCPRC Beijing Office proved time and again the value and necessity of having a local presence in China in carrying out such a study. Jim Reardon-Anderson, former CSCPRC director, should be thanked for believing in the importance of global change issues and in convening a panel to increase our knowledge of Chinese efforts in this area. Beryl Leach, staff officer for the panel, made key contributions to this report: her insights into the organization of

Chinese institutions and extensive knowledge of Chinese research agendas helped guide the panel when it otherwise might have gone astray; her written contributions were valuable to the panel; and her ability to put scientific jargon into a readable form make this report what it is. As chairman, I especially valued her motivation for excellence and her humor when the going got tough. The panel was also assisted by contributions from Kathleen Norman, project assistant for the panel during part of its work, who ably researched and compiled the information contained in Appendix C and who organized the administrative and logistical details of panel meetings. Extensive and helpful literature reviews were carried out Fan Songmiao, Liang Jinyou, Fu Jimeng, and Kate LeJeune. Marc Abramson, CSCPRC summer intern, conducted background research on Chinese global change efforts that helped launch initial panel efforts. I would also like to thank Alice Hogan, program manager for the U.S.-China Program at the National Science Foundation, for the funding from the Division of International Programs that supported the panel's work. Although space does not permit thanking individuals by name, we greatly appreciated the cooperation of the international program officers in U.S. government agencies who kindly provided the panel with information about their agencies' work with China.

This report provides a road map through the Chinese global change program and identifies opportunities for collaborative research. The report will be a success if collaboration increases between the Chinese and U.S. global change communities, and I have no doubt that great potential exists for such an increase. However, global change research requires a change in how collaborative research is supported. As this report shows, most collaborative projects to date are short-term, and by their nature, produce primarily short-term benefits. To obtain the depth of understanding required for global-scale issues, sustained support is required. It is critical that mechanisms be developed to support such efforts in order to take advantage of the opportunities offered to the world by China and its scientists in the arena of global change research.

James N. Galloway, *Chairman*
Panel on Global Climate Change
Sciences in China
Charlottesville, Virginia
October 1992

# Contents

EXECUTIVE SUMMARY ............................................................. 1

1 INTRODUCTION ................................................................ 15
   China's Role in Global Environmental Change, 15
   Purpose and Structure of the Study, 16

2 CHINA'S RESPONSES TO GLOBAL CHANGE ........................... 23
   China's View of Global Change, 23
   Chinese Global Change Program, 24

3 OVERVIEW OF INSTITUTIONS RELEVANT TO
   GLOBAL CHANGE RESEARCH ............................................. 37
   Organization, 37
   Funding, 39
   Chinese Academy of Sciences, 40
   National Environmental Protection Agency, 42
   National Natural Science Foundation of China, 43
   State Education Commission, 45
   State Environmental Protection Commission, 49
   State Meteorological Administration, 49
   State Oceanographic Administration, 49
   State Planning Commission, 50
   State Science and Technology Commission, 51

4   CHINESE PARTICIPATION IN INTERNATIONAL
    GLOBAL CHANGE RESEARCH PROGRAMS ............................... 53
    Introduction, 53
    International Global Atmospheric Chemistry
        Project, 56
    Past Global Changes, 59
    Global Change and Terrestrial Ecosystems, 62
    Biospheric Aspects of the Hydrological Cycle and
        Global Energy and Water Cycle Experiment, 65
    Programs on Marine Environments, 69
    Global Analysis, Interpretation, and Modeling, 78
    System for Analysis, Research, and Training, 79
    Data and Information Systems for the IGBP, 81
    Human Dimensions of Global Environmental
        Change Program, 82
    Chinese Ecological Research Network, 83

5   SELECTED TOPICS .............................................................. 91
    Introduction, 91
    Atmospheric Chemistry, 92
      Trace Gases and Oxidants, 93
      Aerosols, 95
      Stratospheric Ozone, 96
      Atmospheric Deposition, 96
    Physical and Ecological Interactions of
        the Atmosphere and Land Surface, 98
      Hydrology, 98
      Biotic Controls on Trace Gases, 102
      Climate Change Effects on Land Cover Change
        Dynamics, 108

6   SUMMARY .......................................................................... 119
    Introduction, 119
    Contributions to International Research Programs, 121
    Prospects for Collaboration, 123

REFERENCES ............................................................................ 125

APPENDIXES
A   Overviews of Selected Institutions ................................. 135
    Introduction, 135
    Beijing Normal University, 135
    China Remote Sensing Satellite Ground Station, 136

Chinese Academy of Meteorological Sciences, 138
Commission for Integrated Survey of Natural Resources, 139
Guangzhou Institute of Geography, 141
Institute of Atmospheric Physics, 141
Institute of Botany, 143
Institute of Geography, 144
Lanzhou Institute of Glaciology and Geocryology, 147
Lanzhou Institute of Plateau Atmospheric Physics, 150
Nanjing Institute of Environmental Science, 152
Nanjing Institute of Geography and Limnology, 152
Nanjing Institute of Soil Science, 154
Nanjing University, 155
Northwest Institute of Soil and Water Conservation, 156
Peking University, 157
Qingdao Institute of Oceanology, 159
Research Center for Eco-Environmental Sciences, 160
Shanghai Institute of Plant Physiology, 163
South China Institute of Botany, 165
South China Sea Institute of Oceanology, 167
Xi'an Laboratory of Loess and Quaternary Geology, 168
Xinjiang Institute of Biology, Pedology, and Desert Research, 170
Zhongshan University, 171
B  Global Change Projects Listed by the National Natural Science Foundation of China ............................................................. 173
C  Selected Bilateral and Multilateral Global Change Projects ........................................................................ 181
D  Ecological Stations of the Chinese Academy of Sciences ............................................................................. 197
E  Contact Information for Selected Institutions ............................. 203
F  Abbreviations and Acronyms ....................................................... 209

## MAPS

4-1 Sino–Japanese Atmosphere–Land Surface Processes Experiment

## FIGURES

2-1 Linkages among biological, chemical, and physical processes critical to the understanding of global change on a decade-to-century time scale
2-2 Organization of the Chinese National Committee for the IGBP
2-3 Organization of the Chinese national climate research program

3-1 Simplified organization of institutions involved in Chinese global change research, policy, or funding

**TABLES**

3-1 Global change project funding from the National Natural Science Foundation of China, 1986–1992
3-2 Project awards in the Department of Earth Sciences, National Natural Science Foundation of China, 1987–1989

# Executive Summary

## INTRODUCTION

Because of its tremendous population, economic development strategies, and natural resource base, China is causing significant environmental change, with impacts that extend regionally and, in some cases, globally. China will remain heavily dependent on coal to fuel the advances envisioned in its ambitious economic development plans for the 1990s. No doubt exists that anthropogenic emissions will increase. As maximum economic growth policies proceed, land use changes will be greatly accelerated, which have implications for land use patterns, water resources, and atmospheric composition. Due to energy inefficiencies, resource consumption patterns, and increased fertilizer applications, China will alter the regional and global atmospheric chemical composition due to increased trace gas fluxes.

As in other countries, the issues of global environmental change have emerged from the scientific community. As a result, a policy approach to global change issues and research is evolving. China has been forceful in international fora in advocating that wealthy industrialized nations help finance developing countries' participation in regimes addressing global warming. In 1991, China signed the Montreal Protocol on Substances that Deplete the Ozone and is researching and developing chlorofluorocarbon (CFC)-alternative technologies. Governmental support for global change scientific research is emerging.

## Purpose

Given China's current and potential impacts on the global environment and the contributions Chinese science can make to global change research, it is all the more important for China to participate fully in international research programs that address global change questions. However, not much detailed information has been available to program planners or foreign researchers interested in collaboration. Consequently, the CSCPRC requested funding from the Division of International Programs at the National Science Foundation to conduct a study that would report systematically and in greater detail about the organization of Chinese global change science and research activities.

The thrust of the report is twofold. First, and primarily, the report is a reference for individuals who wish to develop collaborative projects with Chinese colleagues, particularly for those who have limited experience in conducting cooperative science in China. To meet this goal, the panel worked hard to find out substantive details about research, despite the limits of available documentation. Secondly, by discussing the way Chinese science is organized, the report provides insights into research priorities, institutional infrastructure, human resources, and other factors that constrain or facilitate Chinese responses to global change.

## CHINA'S VIEW OF GLOBAL CHANGE

China is a good example of how nations respond to this global research agenda from the point of view of their own national interests. According to Ye Duzheng, chairman of the Chinese National Committee for the International Geosphere–Biosphere Program (CNCIGBP), Chinese research on global change will have a definite national focus. From the Chinese viewpoint, "global" change is too large a scale for their needs and current scientific and financial capacities. Specifically, China, like most countries, is very concerned about the possible impact of climate change on economic development and on existing problems such as deforestation, soil erosion, and soil degradation.

Besides emphasizing the regional and local impacts of putative global environmental change, Chinese research also emphasizes studies of historical change and studies of land use problems that are ubiquitous both in China and globally. Studies of phenomena that impact the global environment—such as biogenic and industrial emissions—are apparently of lower priority. Data are not collected or presented systematically and, in some cases, are not made available for proprietary or policy reasons. The Chinese program makes its principal contribution to the international program through analysis of large-scale biophysical phenomena within China and also through analysis of historical changes in China's environment that reflect global and local changes.

Chinese global change research priorities focus on the question of what will be the impact of global change on China. The reverse question should be mentioned: what will be the contribution of China to global change? Although China's focus is practical given its population growth, current and projected industrial base, demands for fossil fuel, and rate of economic development, China's impact on global change is important to the international community. Even though China's basic global change research is expected to remain locally and regionally focused, important opportunities for international collaboration exist that would increase China's and the international community's understanding of the causes and consequences of global environmental change.

## CHINESE GLOBAL CHANGE PROGRAM

Chapter 2 reports on Chinese involvement in two major international global change programs: (1) the International Geosphere-Biosphere Program (IGBP), sponsored by the International Council of Scientific Unions (ICSU), which is devoted mainly to biological and chemical aspects of global change and (2) the World Climate Research Program (WCRP), jointly sponsored by ICSU and the World Meteorological Organization (WMO), which is devoted primarily to physical aspects of global change. While these two programs do not represent all global change research programs, their significance and the fact that China has established national committees for each of them makes these two programs the main focus in this report.

### Chinese National Committee for the IGBP

The Chinese National Committee for the IGBP (CNCIGBP) was established in 1988. Its mission is to organize and coordinate scientists and research communities in China in the study of global change.

According to Ye Duzheng, research at the Chinese Academy of Sciences (CAS), which plays a leading role in global change research, is being organized into three priority areas: (1) attention to sensitive zones and early detection of strong signals of global change (historical, present, and future); (2) human impacts on the environment and, as landscape changes, the effect on trace gas emissions; and (3) use of proxy data in the IGBP core project on Past Global Changes (PAGES).

Under the Eighth 5-Year Plan (1991–1995), the cornerstone of the Chinese global change program is a national key project to study changes in the life-supporting environment in the next 20 to 50 years. (The life-supporting environment is a Chinese term that is defined as the composition of four elements: atmosphere, terrestrial water, vegetation, and soil.)

## Chinese National Climate Committee

In 1987, the State Science and Technology Commission (SSTC) established the Chinese National Climate Committee. The Chinese national climate program consists of five subprograms, which parallel WMO climate programs: (1) the national climate data subprogram, located at the State Meteorological Administration (SMA), is concerned with collecting compatible national data sets and improving monitoring; (2) the climate research subprogram, located at the CAS Institute of Atmospheric Physics, is concerned with modeling, numerical simulation, and observational programs; (3) the Tropical Oceans Global Atmosphere (TOGA) subprogram, located at the State Oceanographic Administration (SOA), is concerned with data and modeling describing the coupling between ocean and atmosphere in the tropics; (4) the national climate application subprogram, located at the Chinese Academy of Meteorological Sciences, is concerned with the use of climate resources; and (5) the national climate impact subprogram, located at the Chinese Research Academy of Environmental Sciences, is concerned with the effects of climate variation and change.

## ORGANIZATION OF CHINESE SCIENCE

The way science is organized and conducted in China has implications for Chinese global change research and for those individuals or foreign institutions that may wish to collaborate. Institutions are vertically integrated and the lack of internal and external disciplinary or programmatic integration can lead to duplication of effort, and problems communicating data across institutional structures limits the effectiveness of research. The problem of parallel vertical organization and resulting lack of integration is especially relevant for global change research, given the need for multi- and interdisciplinary research programs. Chinese scientists are aware of these inefficiencies and compensate through ingenuity and individual collaborations.

Chapter 3 reports on the basic organization of major institutions that conduct research, fund, or set policies that bear significantly on the conduct of global change science in China. CAS has a primary role in basic research and has a large multidisciplinary infrastructure. The National Environmental Protection Agency (NEPA) plays a regulatory role and works on global warming issues and alternative technologies. The National Natural Science Foundation of China (NSFC) funds relevant research. The State Education Commission has a potential role in meeting the educational needs of global change science. The State Environmental Protection Commission has a leading role in policy making. SMA is responsible for meteorological research. SOA is responsible for Chinese ocean areas. The State Planning Commission plays a primary role in setting the 5-year plans. SSTC coordinates and administers civilian science and technology efforts.

## CHINESE PARTICIPATION IN INTERNATIONAL GLOBAL CHANGE PROGRAMS

Chapter 4 presents a broad overview of Chinese activities in IGBP, WCRP, and the Human Dimensions of Global Environmental Change (HD/GEC) Program that is sponsored by the International Social Science Council. Research highlights are presented in each of the core project areas in which China is, plans to be, or has the potential (in the panel's view) to be actively engaged. Also included is a section on the Chinese Ecological Research Network (CERN) because it is a component of the CNCIGBP's global change program. Further details about the organization and research of selected institutions identified in this chapter are provided in Appendix A.

### International Global Atmospheric Chemistry Project

Atmospheric chemistry research is carried out in a number of institutes and universities, usually addressing urban air pollution issues such as oxidants, suspended particles, and toxic species. Recently, some attention has been directed at research projects that have regional and global implications. Most of these projects are closely related to research activities in the International Global Atmospheric Chemistry project. A major focus is on greenhouse gas emissions, including $CH_4$, $N_2O$, and $CO_2$. Research projects on stratospheric ozone have also been carried out. Regional-scale research activities are focused on acid precipitation and oxidants. In addition, the interesting problem of long-range transport of Asian dust and its impact on the Pacific basin has also drawn some attention.

## Past Global Changes

Research on historical analysis of environmental change is voluminous in China. Every CAS institute surveyed lists some form of historical analysis and NSFC has funded this area extensively. Research on historical analysis was nearly ubiquitous among the institutions visited for this report. Every aspect of PAGES research described in IGBP Report No. 12 (1990) or in Bradley (1991) is being reported in China. In fact, the literature is so enormous that it would require a separate and extensive inquiry to catalogue and review materials cited by the Chinese. With few exceptions, work identified by the panel was restricted to China and connections with the rest of the earth system such as telecommunications with global climate anomalies remain to be made. This pervasive consciousness of the variable past has elevated paleoenvironmental studies to a much higher relative level of priority in the Chinese global change program than in the United States or Europe.

This research also demonstrates the unique paleoenvironmental records in China that are a crucial resource for the global PAGES community. The quality of scholarship and skill in this area seemed very high. Integration of prospective research (designed to provide projections of future changes and give prescribed scenarios such as climate and land use changes) into Chinese historical analysis will further enhance their contributions in this area.

## Global Change and Terrestrial Ecosystems

Leading research relevant to the Global Change and Terrestrial Ecosystems core project is being conducted at two CAS institutes. Work at the Institute of Botany provides a very strong foundation for studies of the effects of climate and $CO_2$ change on terrestrial vegetation. Extensive and sophisticated work on agricultural ecosystems was identified at the Shanghai Institute of Plant Physiology that provides very valuable foundations for further studies.

## Biospheric Aspects of the Hydrological Cycle and Global Energy and Water Cycle Experiment

Chinese researchers are active in modeling and measuring interactions between the land surface and the atmosphere. Significant efforts are ongoing in developing models to represent the role of vegetation in controlling surface energy balance and evapotranspiration. The landscape of western China contains considerable contrast between well-watered irrigation districts along its rivers and surrounding arid areas. These contrasts produce effects on mesoscale atmospheric circulation and this phenomenon is the subject of both theoretical and empirical study, such as the Sino-Japanese Atmosphere–Land Surface Processes Experiment. Finally, because of China's historical dependence upon irrigation, hydrology is a mammoth enterprise, and considerable data on surface and groundwater hydrology exists at all scales, though these data are not—as yet—well integrated into the global change endeavor. Overall, China is poised to conduct significant work under the joint IGBP–WCRP Biospheric Aspects of the Hydrological Cycle and Global Energy and Water Cycle Experiment.

## Programs on Marine Environments

With an 18,000 km coastline, China has considerable reason to be interested in the ocean, particularly the coastal zone. Historical, contemporary, and future changes in land cover and basin hydrology will continue to alter drastically the delivery of water and sediments as well as eolian materials to the coastal zone, particularly in the deltaic regions of the Huanghe, Yangtze, and Pearl Rivers. These changes pose serious threats to the livelihoods of millions of people.

The panel's review of Chinese research in the Joint Global Ocean Flux project showed that much of it would be more appropriately considered under the proposed IGBP core project on Land–Ocean Interactions in the Coastal Zone (LOICZ). TOGA involvements are ambitious and valuable components of IGBP and WCRP. Participating scientists are well qualified and eager for international cooperation. In addition to marine research, China offers a combination of land-based research endeavors that are very important to a LOICZ focus on land-ocean interactions. It is quite likely that Chinese efforts would be restricted to Chinese coastal zones—whether in cooperation with TOGA or LOICZ—unless activities were funded from international sources. As LOICZ planning continues, involvement of Chinese scientific leadership might bring particular strength to activities in coastal Asia—an area identified by LOICZ planners as being especially interesting and important from a global perspective.

## Global Analysis, Interpretation, and Modeling

The Chinese IGBP effort places relatively little emphasis on the central questions of the Global Analysis, Interpretation, and Modeling (GAIM) initiative. However, many Chinese activities contribute materially to it. For example, this report describes work on ecosystem dynamics, land process modeling, marine modeling, and climate modeling. As GAIM evolves internationally, it is likely that Chinese participation will increase, and involvement of Chinese scientists should be encouraged to allow them to build upon activities already under way.

### System for Analysis, Research, and Training

The CNCIGBP is developing a proposal to the System for Analysis, Research, and Training Standing Committee for the establishment of a regional research network for East Asia and Western Pacific that would have a regional research center in China. The draft proposal demonstrates the large and multidisciplinary research enterprise that CAS offers the study of global change.

### Data and Information Systems for the IGBP

Since its inception, the IGBP Data and Information Systems supporting project has been open to China's active involvement, for example, in the development of an advanced very high resolution radiometer (AVHRR) 1 km global data set. Researchers at the CAS Institute of Atmospheric Physics, in collaboration with the State Meteorological Satellite Center, will produce 1 km AVHRR data sets for the first time in China.

### Human Dimensions of Global Environmental Change Program

According to NSFC, China plans to participate in the Human Dimensions of Global Environmental Change (HD/GEC) Program. The goals of the HD/GEC program are consistent with China's research priorities; in fact, the study of the impact of human activities is evident to one degree or another in many current research agendas. However, this area of the global change research agenda often requires interdisciplinary research, for which China will have to make certain changes in the way research currently is organized.

### Chinese Ecological Research Network

Over the next five years, CAS will implement the Chinese Ecological Research Network (CERN) by upgrading 30 of its 52 ecological research and monitoring stations. CERN is an ambitious and important commitment by CAS to improve the way it conducts ecological research. The imposing information and data management requirements will demand closer attention to quality assurance and quality control, documentation, and standards. The scope of this undertaking has significant implications for the types of contributions China can make to ecological studies and international scientific programs.

## SELECTED TOPICS

In Chapter 5, the panel addresses two focal areas: atmospheric chemistry and physical and ecological interactions of the atmosphere and land surface. Within these areas, specific topics were chosen by the panel based on members' expertise and panel opinion about the relevance to U.S. and international research. It should be noted that no attempt was made to be comprehensive in examining all of the possible topics available for discussion in a given focal area.

### Atmospheric Chemistry

The major energy source in China is coal. Emissions of particulate matter and $SO_2$ from burning coal are major contributors to regional air pollution. These emissions not only contribute to urban and regional pollution problems such as oxidants and acid precipitation, but potentially also have global impacts. Remarkably high levels of tropospheric $O_3$ over northeastern China and Japan in spring and summer have been deduced from satellite observations.

Chinese atmospheric chemistry research has been conducted primarily in areas of urban pollution, for example, suspended particles, $O_3$ and $O_3$ precursors, and toxic species. Recently, there have been some important efforts to address large-scale background atmospheric chemistry issues that have regional or global implications. The major foci of these efforts include tropospheric oxidants, greenhouse gases, aerosols, stratospheric $O_3$, and acid precipitation. However, these efforts are severely limited due to a lack of funding, advanced instruments, and expertise in a few global change-related disciplines. It appears that atmospheric chemistry is not a field of high priority.

#### *Trace Gases and Oxidants*

Research on trace gases other than urban air pollutants started in recent years when it was realized that all of the trace gases other than $CO_2$ contribute equally as much as it does to climate change. Much of the attention has been on $CH_4$, $N_2O$, and $CO_2$ emissions from various biogenic sources such as rice paddies and forests.

Like many cities in the world, high levels of $O_3$ are a major air pollution problem in most of the urban areas in China. While $O_3$ concentrations are routinely monitored over urban areas by local NEPA bureaux, few observations are being made in remote regions. An exception was the shipboard measurements of atmospheric $O_3$ over the western Pacific Ocean.

*Aerosols*

Current research on aerosol chemistry in China is more limited than on other areas of atmospheric chemistry. Aerosol studies focus primarily on urban- and regional-scale problems. Only a few studies directly address global aerosol distributions and trends or link aerosols to climate change. Based on the information available to the panel, it appears that the importance of aerosols to climate change is not generally appreciated by researchers in China.

Wind-blown dust is believed to contribute significantly to particulate loading, especially in northern China. Aerosol measurements over China, Japan, and the northern Pacific have convincingly demonstrated that dust storms originating from central Asia are the major sources of dust, sulfate, nitrate, and other particulate matter transported to the northern Pacific.

Given the important role played by aerosol particles in atmospheric radiation, the effect of Asian dust storms on regional—as well as global—climate needs to be carefully studied. While China has programs to study the meteorological characteristics of dust storms, including the formation and transport of the storm's dust, a comprehensive program that addresses both chemical and physical properties of dust storms would be welcome.

*Stratospheric $O_3$*

At least four Chinese institutions are engaged in the development of one- and two-dimensional models for stratospheric chemistry studies. Because most of the stratospheric observations and laboratory measurements are carried out in the United States and Europe, Chinese modelers do not often have timely access to these data sets. Computer facilities are also somewhat inadequate to run fully coupled two-dimensional transport and chemical models efficiently. As a result, stratospheric models in China are not as advanced as those in developed countries. In particular, lack of access to observational data is a serious limitation for the development of Chinese stratospheric models. Chinese researchers are measuring $O_3$, $NO_2$, and the consumption of halons and CFCs.

EXECUTIVE SUMMARY 11

*Atmospheric Deposition*

National and regional programs on the measurement of precipitation chemistry are being carried out. Several institutes in China have the skilled personnel to make state-of-the-art determinations of precipitation composition and have done so on a regional basis. The data sets from these studies, especially the more recent ones, are of high quality and the publications resulting from those data are of interest to the global community. Unfortunately, the precipitation composition data available at this time generally are not adequate to address the question of China's impact on global change. The most intractable problem is one of data availability. If Chinese agencies do not provide open access to their data by scientists both within and outside China, then questions that require the use of precipitation data cannot be adequately answered.

## Physical and Ecological Interactions of the Atmosphere and Land Surface

### *Hydrology*

Because of its vital importance to agriculture and economic development, water is considered by the Chinese to be the most precious natural resource. China's approach to studying the hydrological cycle probably will remain very focused on Chinese resource management issues and on social and economic impacts. The relationship between climate change and hydrology appears to be a priority, as evidenced by the work of the National Climate Change Coordination Group and by current and planned research by major institutions such as the Ministry of Water Resources and CAS.

The panel identified research that focuses on two problems of water resource management. The first problem involves increased water pollution due to increasing population, urbanization, and demand for industrial and agricultural outputs. The second problem concerns the uneven distribution of water resources, since most of the water supply is concentrated in the southern part of China while the northern areas have experienced increasing levels of drought in recent years.

### *Biotic Controls on Trace Gases*

The Chinese effort in biogeochemistry has numerous components relevant to land-atmosphere interactions. The panel reports on several that were observed in some depth. First, a program measuring

$CH_4$ emissions from rice cultivation is in progress, conducted as bilateral collaborations between CAS and the Fraunhofer Institute in Germany and between CAS and the U.S. Department of Energy. Second, the Chinese have initiated efforts to examine $N_2O$ production and $CH_4$ consumption from upland soils at the CAS Research Center for Eco-environmental Sciences. Third, an ambitious program of vegetation dynamics analysis at the CAS Institute of Botany supports national and regional estimates of carbon storage and primary productivity. Researchers are also conducting some studies of soil nutrients and organic matter.

The barriers to interdisciplinary study of biogeochemistry in China are clear, given the disciplinary nature of the basic research and funding organizations. In addition, state-of-the-art research in biogeochemistry requires access to instrumentation, reliable analytical standards, and field site travel. All of these requirements can be constraining in China. However, many activities are quite vigorous and international collaboration is strong. The potential for increased collaboration seems high, and Chinese scientists are eager for enhanced activity in this area.

### Climate Change Effects on Land Cover Change Dynamics

Research on land cover and land use change—particularly on historical change—is profuse. More recently, literature has begun to emerge that treats land cover change as part of global change science, although it is no more than regional in scope. Desertification is a major research area in China.

Land use is a prominent issue in natural resources research and in the planning of China's national global change program. Contemporary and planned research on land cover change in China is fragmented and largely historical in approach. Research currently is a national program on phenomena driven, in part, by global mechanisms. Much of the work under way is in the category of historical analysis. Many components are available to develop a focused land cover change program in China: maps, remote sensing, historical records, and some modeling capacity. This is an area in which some catalytic action through international collaboration could make a big difference.

## SUMMARY

Chapter 6 summarizes the panel's major findings concerning China's contributions to global change research and prospects for collaboration. The findings are not comprehensive, nor are recommendations made that would overreach the information the panel was able to collect. The panel concentrated on two areas: enhancing Chinese contributions to international research programs and enhancing bilateral collaborative science.

While problems exist in the Chinese global change program and in the way science is organized in China, it should be pointed out that the Chinese are in good company. More importantly, the Chinese are already making significant contributions to global change research and have the potential for even greater contributions.

Increasing collaboration with China in global change research would have direct benefits for all parties concerned. However, it is necessary to be realistic about how much of the potential can be turned into reality. As described in the report, many difficulties remain in conducting science in China, despite the many substantive improvements the Chinese have instituted since the early 1980s. In general, funding for science in China remains very limited and governmental support will likely remain driven by domestic priorities. It is reasonable to conclude that collaborative global change research and participation in international global change research programs will be highly dependent upon funding from bilateral and multilateral sources.

Developing cooperative science agendas with China in any area is time consuming and requires much more than just a good idea for research. Limited funding and the way Chinese science is organized can combine to frustrate bilateral projects.

The panel found strong potential for increased cooperation in all areas it examined. But, the development of actual cooperation most likely will require the involvement of personnel experienced in cooperative science projects in China and substantive resources to develop and administer the projects. These conditions will be especially important in larger scale projects and in any project requiring access to tightly controlled or fragmented sources of data.

# 1

# Introduction

## CHINA'S ROLE IN GLOBAL ENVIRONMENTAL CHANGE

Because of its tremendous population, economic development strategies, and natural resource base, China is causing significant environmental change, with impacts that extend regionally and, in some cases, globally. The Chinese population stands at more than 1.1 billion people and it is increasing by more than 15 million annually—a figure equal to the total population of Australia. China is pursuing economic policies designed to achieve rapid growth industrially and in agricultural production. This growth will be fueled by high-sulfur coal, which accounts (and is expected to continue to account) for approximately three-quarters of China's annual energy consumption. As a result, 20 million tons of coal dust and 15 million tons of sulfur dioxide are emitted each year. The emission of biogenic methane from agricultural practices, specifically rice paddy production, is significant.[1] The large deserts in western China are also important to the global environment; their dust—mixed with industrial pollutants—is transported over the Pacific Ocean, which may alter the chemistry and radiative processes in the remote troposphere.

To the extent that environmental stresses threaten domestic economic development and food production, the Chinese government has been responsive. Furthermore, environmental pollution and the degradation of natural resources are substantive problems that have caught the public's attention. But, like the rest of the world, the concept of global change is neither well understood nor are the effects of global change tangibly felt in the lives of ordinary citizens. Consequently, global change is not top policy priority when matched against the readily apparent consequences of China's environmental problems. That is not to say that no interest or concern exists. The impact of global warming, for example, has been addressed at the national level. In the "National Report of the People's Republic of China on Environment and Development" (SPC 1991), the important implications of global warming on agricultural output and sea-level rise were explicitly noted. Moreover, the Asian Development Bank has recently approved a $600,000 technical assistance grant to China (to be administered through the State Science and Technology Commission [Chapter 3]) to formulate a national response strategy for global climate change.

Despite a clear acknowledgement of environmental problems and calls for substantive mediation, a national policy on the issues of global environmental change is not well formed. In China, as in other countries, the issues of global environmental change have emerged from the scientific community. And, the Chinese scientific community is responding, assisted, in part, by major international research programs that address global climate and global environmental change.

As a result, a policy approach to global change issues and support for research is evolving. China has been forceful in international fora in advocating that wealthy industrialized nations help finance developing countries' participation in regimes addressing global warming. In 1991, China signed the Montreal Protocol on Substances that Deplete the Ozone (herein referred to as the Montreal Protocol) and is researching and developing chlorofluorocarbon (CFC)-alternative technologies.

China will remain heavily dependent on coal to fuel the advances envisioned in its ambitious economic development plans for the 1990s. No doubt exists that anthropogenic emissions will increase. As maximum economic growth policies proceed, land use changes will be greatly accelerated, which have implications for land use patterns, water resources, and atmospheric composition. Also, due to energy inefficiencies, resource consumption patterns, and increased fertilizer applications, China will alter the regional and global atmospheric chemical composition due to increased trace gas fluxes.

## PURPOSE AND STRUCTURE OF THE STUDY

### Background

The National Academy of Sciences (NAS), through the Committee on Scholarly Communication with the People's Republic of China

(CSCPRC), has been working for a number of years to promote China's full participation in international global change research programs, in particular the International Geosphere–Biosphere Program (IGBP). In 1987, NAS sent John Eddy, James McCarthy, and Harold Mooney, three leading figures in the U.S. global change science community, to China to discuss global change and possible areas for collaboration. Based on that visit, the National Science Foundation (NSF) provided funding to the CSCPRC to foster scientific collaboration.[2]

Even though these individual collaborations were being organized with some success by CSCPRC and by researchers themselves, it became increasingly apparent that Western scientists lacked any broadly based or detailed knowledge about the development and status of Chinese national responses to global change. It was not that the Chinese were trying to withhold information; on the contrary, they were disseminating lists of projects in research areas under the IGBP, for example. However, this information lacked details crucial to any substantive understanding of the organization of the Chinese global change programs and of the contents of research projects. Documentation in English usually was limited to project titles; details about project design, objectives, implementation period, or principal investigators and institutes usually were not provided.

## Purpose

Given China's current and potential impacts on the global environment and the contributions Chinese science can make to global change research, it is all the more important for China to participate fully in international research programs that address global change questions. However, not much detailed information has been available to program planners or foreign researchers interested in collaboration. Consequently, the CSCPRC requested funding from the Division of International Programs at NSF to conduct a study that would report systematically and in greater detail about the organization of Chinese global change science and research activities.

The thrust of the report is twofold. First, and primarily, the report is a reference for individuals who wish to develop collaborative projects with Chinese colleagues, particularly for those who have little or no direct experience in conducting cooperative science in China. To meet this goal, the panel worked hard to find out substantive details about research, despite the limits of available documentation. Secondly, by discussing the way Chinese science is organized, the report provides insights into research priorities, institutional infrastructure, human resources, and other factors that constrain or facilitate Chinese responses to global change.

## Committee Charge

With NSF support, the CSCPRC formed a panel of U.S. experts to study and report on the state of basic research in China in the area of climate studies and global change and to give detailed description and analysis of a selected number of topics. The panel was asked to provide both broad and in-depth coverage of the state of global change sciences in China. First, the overview would include: (1) the organization, composition, and agenda of the Chinese National Committee for the IGBP (CNCIGBP) and the Chinese National Climate Committee; (2) projects, institutions, personnel, research, education, centers of excellence, international cooperation (with special emphasis on Pacific Asia), and other scientific activities related to global change; (3) science policies, administration, funding, and research priorities; and (4) the larger social and economic environment that bears on the global change agenda.

To substantiate the overview further, the panel was asked to focus secondly and in greater detail on a selected number of disciplines or research topics that complement specific components of the U.S. national research program and the IGBP core projects. These focal assessments would illuminate capabilities and policy commitments and explore the potential for collaboration with the U.S. scientific community and the potential for contributions to international research. The panel chose two focal areas: atmospheric chemistry and physical and ecological interactions of the atmosphere and land surface. In the first area, the panel examined oxidants, trace gases, aerosols, and atmospheric deposition. And, in the second area, the panel examined hydrology, biotic controls on trace gases, and land cover change dynamics.

Based on this charge, the panel was formally constituted as the "CSCPRC Panel on Global Climate Change Sciences in China." When the panel met together for the first time, however, members noted that the committee charge went beyond climate change studies and, therefore, approached its work under the broader rubric of reporting on global change sciences.

## Methodology

The panel met in Washington, D.C. in March 1991 for an organizational meeting to determine the methodology for the study and to make research, writing, and travel assignments. Literature reviews were commissioned to assist panel members in their report writing. Focal topics outlined above were chosen based on panel members'

INTRODUCTION                                                                  19

expertise and areas of interest, and no attempt was made to be comprehensive, even within a focal area. In order to collect detailed information about research activities and institutional capabilities, panel members decided to spend various amounts of time in China.

The panel enlisted the cooperation of the CNCIGBP in compiling English-language summary documents and in organizing seminar presentations on research. CNCIGBP is administered by the Chinese Academy of Sciences (CAS), which is the largest institution in China carrying out research across the range of global change disciplines. Because CAS plays a primary role in global change research and because of its role in organizing panel access to information, the study's information about CAS activities is particularly robust compared to other agencies.

In June 1991, panel members James Galloway, Joseph Berry, C.S. Kiang, Shaw Liu, David Schimel, Wei-Chyung Wang, John Winchester, and Beryl Leach, panel staff officer, visited China as a group. Under the direction of Ye Duzheng, chairman of the CNCIGBP, materials were collected for the panel, itineraries for panel members were arranged, and a two-day seminar was held at which agencies and CAS institutes were able to introduce their current and planned global change research. Nien Dak Sze visited China in July 1991. For approximately 2 weeks in August, William Reiners visited Beijing, Nanjing, Shanghai, and Guangzhou.

Even though the panel members recognized that time and financial constraints would prevent a comprehensive itinerary of visits to institutes engaged in global change research, visits were recognized as a valuable and necessary way to collect the type of detailed information the panel sought. Using a survey tool devised after the first panel meeting, members surveyed each institute for a consistent set of details about organization, research projects, equipment, and research facilities in order to identify centers of excellence and potential areas for collaboration. Visits to institutions involved in the management or funding of science were also arranged in order to improve information about the organization of science and current science policy priorities. See Chapter 3 and Appendix A for details.

To compensate for the limited amount of time available to make visits, the panel drew extensively and very successfully on the resources and expertise of the CSCPRC Beijing Office. Office personnel followed up on panel visits and provided substantive support in collecting details to supplement Chinese documentation and the panel's preliminary findings. Use of this office was crucial to the panel's successful completion of its charge.

In November 1991, most panel members met in Washington D.C.

to analyze and integrate their findings and draft the report. Subsequently, small *ad hoc* panel meetings were held in Baltimore, Maryland (two), Boulder, Colorado (one), and Charlottesville, Virginia (three) to work on report sections.

Because of the scope of the study and the limited time available for research in China, the panel recognizes that its findings are not exhaustive. It will rely on the efforts of others to help complete the picture now being sketched in this report.

## Report Format

Chapter 1 describes why it is important to understand the role China plays in global change and China's need to be fully integrated into the major international global change research programs, which provide the justification for the study. In addition, Chapter 1 describes the terms, methodology, and format for the study and report. Chapter 2 describes how China views global change and how it has organized its research efforts by forming national committees for the IGBP and the World Climate Research Program (WCRP). Chapter 3 provides an overview of the major institutions involved in research, policy, or funding of global change research. Chapter 4 describes Chinese research that is relevant to IGBP and WCRP core programs and the Chinese Ecological Research Network, which is considered an official component of the Chinese global change program. Chapter 5 focuses in greater detail on specific research topics selected by the panel. Chapter 6 presents the panel's major findings concerning China's contributions to global change research and prospects for collaboration. Appendix A contains detailed reports from institutions visited by panel members. Appendix B presents a listing for the National Natural Science Foundation of China of recent Chinese global change projects for which it provides either partial or total funding. Appendix C lists selected global change research being conducted under the U.S.-China scientific protocols, relevant university-level bilateral research known to the panel, and projects in which China is involved with other Pacific Rim or European countries. Appendix D lists CAS ecological research stations, including those in CERN and the CERN plan for major ecosystem research. Appendix E provides contact information for investigators and institutions referenced in this report. Appendix F lists abbreviations and acronyms used in the report.

## NOTES

1. Some of these gases and what happens to them in photochemical reactions are also important, as infrared and reactive active species can accumulate in the atmosphere and alter climate and oxidative capacity in the global environment.

2. The *ad hoc* USA-PRC Committee for the Joint Study of Global Change, under the co-chairmanship of James N. Galloway (University of Virginia) and James Gosz (University of New Mexico) was established to facilitate the development of cooperative studies by organizing and funding project development workshops. By 1991, six areas and co-principal investigators had been identified and five bilateral workshops had been held: climate-vegetation interactions, precipitation composition, global change education (proposal pending), air and water transport of soils (two funded proposals), background concentrations of trace gas fluxes to the atmosphere (proposal pending), and building links between the Chinese Ecological Research Network and the U.S. Long-Term Ecological Network (three funded projects).

# 2

# China's Responses to Global Change

## CHINA'S VIEW OF GLOBAL CHANGE

In the international dialogue, global change research addresses the causes and consequences of global environmental change on regional and local systems. Also, global change research may address problems that are local in nature but which are occurring ubiquitously and so have become global problems (such as soil erosion). However, a certain imprecision exists in this definition that can cause differences in determining what research will be considered under the global change rubric. In this light, the panel believed that some amount of research may be going on in China that is relevant, but that it is not being identified at this time as global change research by the Chinese themselves.

China is a good example of how nations respond to this global research agenda from the point of view of their own national interests. According to Ye Duzheng, chairman of the Chinese National Committee for the International Geosphere-Biosphere Program (CNCIGBP), Chinese research on global change will have a definite national focus. From the Chinese viewpoint, "global" change is too large a scale for their needs and current scientific and financial capacities. Hence, the main research objective for the CNCIGBP is to concentrate on areas that are of practical importance for China and of general importance to other countries—in particular, developing countries—and that are at the same time scientifically challenging. Specifically, China, like most countries, is concerned about the possible impact of climate change on economic development and on existing problems such as deforestation, soil erosion, and soil degradation.

Besides emphasizing the regional and local impacts of putative global environmental change, Chinese research also emphasizes studies of historical change and studies of land use problems that are ubiquitous both in China and globally. Studies of phenomena that impact the global environment—such as biogenic and industrial emissions—are apparently of lower priority. Data are not collected or presented systematically and, in some cases, are not made available for proprietary or policy reasons. The Chinese program makes its principal contribution to the international program through analysis of large-scale biophysical phenomena within China and also through analysis of historical changes in China's environment that reflect global and local changes.

Chinese global change research priorities focus on the question of what will be the impact of global change on China. The reverse question should be mentioned: what will be the contribution of China to global change? Although China's focus is practical given its population growth, current and projected industrial base, demands for fossil fuel, and rate of economic development, China's impact on global change is important to the international community. Even though China's basic global change research is expected to remain locally and regionally focused, important opportunities for international collaboration still exist that would increase China's and the international community's understanding of the causes and consequences of global environmental change.

## CHINESE GLOBAL CHANGE PROGRAM

China has been involved in two major international global change programs (Figure 2-1) since their early stages: (1) the International Geosphere–Biosphere Program (IGBP), sponsored by the International Council of Scientific Unions (ICSU), which is devoted mainly to biological and chemical aspects of global change and (2) the World Climate Research Program (WCRP), jointly sponsored by ICSU and the World Meteorological Organization (WMO), which is devoted primarily to physical aspects of global change. Current Chinese plans for global change research have been developed for both of these activities (in contrast to the United States, where scientific committees and federal agencies have worked towards a single program). While these two programs do not represent all global change research programs,

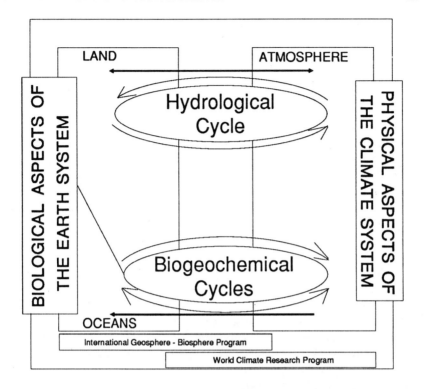

FIGURE 2-1 Linkages among biological, chemical, and physical processes critical to the understanding of global change on a decade-to-century time scale. Arrows refer to the seven first research priority questions described in IGBP Report No. 12 (1990). SOURCE: IGBP. Used with Permission.

their significance and the fact that China has established national committees for each of them makes these two programs the main focus in this report.[1]

## Chinese National Committee for the IGBP

### *Organization and Membership*

The Chinese National Committee for the International Geosphere–Biosphere Program (CNCIGBP) (Figure 2-2) was established in 1988 with the mission to organize and coordinate scientists and research communities in the study of global change. The China Association for Science and Technology (CAST), a nongovernmental umbrella or-

FIGURE 2-2 Organization of the Chinese National Committee for the IGBP (CNCIGBP) (CAS: Chinese Academy of Sciences; CAST: China Association for Science and Technology).

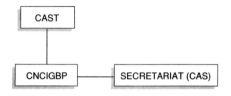

ganization for professional scientific societies, is the official representative of China to ICSU and sponsor of the CNCIGBP. The Chinese Academy of Sciences (CAS) acts as the committee's secretariat, which is located in the CAS Bureau of Resources and Environmental Sciences. It is important to note that the committee does not administer funding for global change research, nor does it have the authority to command a coordinated, centrally planned global change research agenda.

CNCIGBP membership includes all of those individuals who responded to an invitation sent to many organizations. As the global change program has evolved, membership has grown (from 23 in 1991 to 39 in 1992) to reflect a wider range of institutions, administrative systems (for example, ministries, CAS, State Meteorological Administration [SMA], and State Oceanographic Administration [SOA]), and additional leading figures in the Chinese scientific community. CNCIGBP membership is listed below:

> Song Jian (honorary chairman), state councilor; director, State Science and Technology Commission (SSTC)
> *Ye Duzheng (chairman), special advisor, CAS; director *emeritus*, Institute of Atmospheric Physics, CAS
> *Chen Jiaqi, Department of Water Resources, Ministry of Water Resources (MOWR); Chinese Society of Hydraulic Engineering
> *Chen Panqin (secretary general), Bureau of Resources and Environmental Sciences, CAS
> Chen Qinglong, Bureau of Science and Technology, State Education Commission (SEDC)
> Chen Shupeng, Institute of Geography, CAS; Geographical Society of China
> Chen Yongning, Institute of Zoology, CAS; Ecological Society of China
> Cui Haiting, Geography Department, Peking University; Geographical Society of China
> *Deng Nan (vice chairman), vice director, SSTC

---

*also member of the Chinese National Climate Committee.

Ding Yihui, Chinese Academy of Meteorological Sciences (CAMS), SMA; Meteorological Society of China
*Fang Weiqing, Department of Science and Technology for Social Development, SSTC
Fu Congbin, Institute of Atmospheric Physics, CAS, Meteorological Society of China
Guo Shiqin, Ministry of Agriculture (MOA)
Hong Yetang, Guiyang Institute of Geochemistry, CAS; Chinese Society of Mineralogy, Petrology, and Geochemistry
*Hu Dunxin, Ocean Circulation and Air-Sea Interaction Laboratory, Qingdao Institute of Oceanology, CAS; Society of Oceanology and Limnology
Huang Bingwei, Institute of Geography, CAS; Geographical Society of China
Li Wenhua, Commission for Integrated Survey of Natural Resources (CISNAR), CAS; China Research Society of Natural Resources
Li Yongqi, Department of Marine Biology, Qingdao University of Oceanography; Chinese Society of Oceanography
Lin Hai, Department of Earth Sciences, National Natural Science Foundation of China (NSFC)
Liu Tungsheng, director, Xi'an Laboratory for Loess and Quaternary Geology; Chinese Society of Mineralogy, Petrology, and Geochemistry
Liu Shu (vice chairman), CAST
Lu Jingting, Department of International Affairs, CAST
Lu Shouben, SOA; Chinese Society of Oceanography
Ma Yang, Department of Society Activities, CAST
Ouyang Ziyuan, Bureau of Resources and Environmental Sciences, CAS
Shen Qiuxing, MOA; Agricultural Society of China
Shen Shanmin, director, Institute of Applied Ecology, CAS; Soils Society of China
Sun Honglie (vice chairman), vice president, CAS; director, CISNAR, CAS; China Research Society of the Qinghai–Tibet Plateau
Wang Lixian, Beijing Forestry University; Soil and Water Conservation Society of China
Wang Ren (vice chairman), NSFC
Xiao Xuchang, Institute of Geological Science, Ministry of Geology and Mineral Resources (MOGM); China Research Society of the Qinghai–Tibet Plateau
Xu Deying, Institute of Forestry, Chinese Academy of Forestry Science; Forestry Society of China

Yuan Daoxian, Guilin Institute of Karst Geology, MOGM; Geological Society of China
Zhang Shen, Institute of Geography, CAS; Chinese Society of Environmental Science
Zhang Xinshi, Institute of Botany, CAS; Botanical Society of China
Zhang Yue, Department of Rural and Forest Water Resources Protection, MOWR; Soil and Water Conservation Society of China
Zhang Zonghu, Institute of Hydrological Engineering, MOGM; Geological Society of China
Zhao Qiguo, Nanjing Institute of Soil Science, CAS; Soils Society of China

### *CNCIGBP Research Agenda*

By September 1990, based on findings from pilot studies, the CNCIGBP had identified what kinds of global change were most important for China at present and in the near future. From this exercise, the committee reported its initial priorities and contributions to ICSU in the following IGBP core areas:

- International Global Atmospheric Chemistry (IGAC) Core Project
- Past Global Changes (PAGES) Core Project
- Global Change in Terrestrial Ecosystems (GCTE) Core Project
- Biospheric Aspects of the Hydrological Cycle (BAHC) Core Project
- Joint Global Ocean Flux (JGOFS) Core Project
- Land–Ocean Interactions in the Coastal Zone (LOICZ) Core Project (proposed)
- Global Analysis, Interpretation, and Modeling (GAIM) Special Committee
- Regional Research Centers (RRCs)[2]
- Data and Information Systems (DIS) for the IGBP

The Chinese also list the Chinese Ecological Research Network (CERN) as an IGBP activity, and so it will be addressed in more detail in Chapter 4 even though it is not directly subsumed under any of the previously mentioned IGBP core projects.

According to Ye Duzheng, research at CAS is being organized into three priority areas: (1) attention to sensitive zones and early detection of strong signals of global change (historical, present, and future); (2) human impacts on the environment and, as landscape changes, the effect on trace gas emissions; and (3) use of proxy data in the IGBP core project on Past Global Changes.

In May 1991, the CNCIGBP published its first issue of the *Bulletin of the CNCIGBP*, which is written in English and is designed to increase the information available about Chinese global change research (CNCIGBP 1991). The committee plans to publish the bulletin regularly. This publication is useful and will become even more valuable when the committee reports on project details, including the identification of investigators, responsible institutions, funding sources, duration, and project findings.

According to the bulletin, a nine-expert committee, chaired by Ye Duzheng, has been established to coordinate global change research programs in the Eighth 5-Year Plan (1991–1995). This committee will organize five scientific workshops to help implement the Chinese global change research program, and the following workshop topics have been identified:

- Past global change in China. The objective will be to establish the conceptual model of past global change. Scientists will focus on three areas: (1) techniques and methods for extracting historical data and information and reconstruction of climate in China during the past 2,000 years; (2) definition of the possible global events and regional events in paleoclimate and environment series; and (3) studies of dominant factors and key interactive processes in past changes.
- Climate change effects on terrestrial ecosystems. The objective will be to study global change mechanisms and interactions among the Earth's elements. Scientists will focus on two areas: (1) analysis of the effects of climate on terrestrial ecosystems based on various data, especially satellite data; and (2) theoretical analysis and numerical modeling of effects of climate on terrestrial ecosystems.
- Impact of human activities on biological sources of trace gases, the water cycle, and energy exchange. The objective will be to study the nature and extent of the impact of human activities on the life-supporting environment. Scientists will focus on four areas: (1) comparative study of exchange processes of water, heat, and trace gases in natural and man-made ecosystems; (2) estimation of biomass and identification of key ecosystem processes and variables; (3) identification of trace gases produced by human activities; and (4) water cycle processes in the soil–vegetation–atmosphere system under different land uses.
- Sensitive areas of environmental change and detection of early signals of significant global change. The objective will be to seek early strong signals of global change in ecologically sensitive zones. Scientists will focus on four areas: (1) detection techniques and methods; (2) intensive observation in areas sensitive

to climate change; (3) calibration of remote sensing data; and (4) analysis of land–surface features by using airborne remote sensing techniques and satellite images.
- Characteristics and trends of changes of the life-supporting environment in China. The objective will be to study methods of predicting global change in the life-supporting environment. Scientists will focus on three areas: (1) predictability of global change; (2) development of regional models to describe and interpret changes in climate and ecosystems (which are reflected in the data analyses of the previous topics); and (3) development of dynamic models of environmental change. This workshop topic is also a national key project (see below).

Under the Eighth 5-Year Plan, one of the most significant activities under the Chinese global change program is a national key project to study changes in the life-supporting environment in the next 20 to 50 years. (The life-supporting environment is a Chinese term that is defined as the composition of four elements: atmosphere, terrestrial water, vegetation, and soil.) Because the time scale is 50 years, the natural processes leading to soil property changes will not be considered; instead, the Chinese will study only human activities causing those changes. Because natural vegetation change is also a slow process (although quicker than that of soil), research will concentrate on sensitive or transitional zones (sections or zones between different climate regions) and will include the study of human influences on the processes of vegetation changes. The project was selected by SSTC, and Ye Duzheng is the lead scientist.

The CNCIGBP provided the panel with the following list of 32 projects relevant to the IGBP that were developed between 1981 and 1990:[3]

- Study of the air–sea interaction in the tropical west Pacific region and its impact on annual climate change in China
- Sino–Japanese cooperative study of the Kuroshio Ocean Current
- Study of the Chinese Quaternary coastline and forecasting of sea-level and coastline changes
- Study of aerosols over the Antarctic Ocean
- Experimental observation and case study of comprehensive development and administration of eco-environmental resources in typical Chinese regions
- Development of the Loess Plateau and global environmental change
- Study of the content and distribution of carbon, sulfur, and nitrogen in China's lithosphere

- Development and application of remote sensing techniques
- Farming and herding experiments for developing and using a saline lake productively
- Study of the reasons for and forecast of drought and flood in the Yangtze and Huanghe River Basins in China (1988–1992)
- Study of the biogeochemistry of carbon, nitrogen, sulfur, and phosphorus in Haihe River basin in China
- Preliminary study of China's sea level and climatic change: their trends and impacts (1988–1992)
- Comprehensive land restoration experiment on the Sanjiang Plain in northeast China
- Study of regional comprehensive development and land restoration strategies in southwest China
- Prevention and treatment technologies for acid rain
- Study of environmental background concentrations and capacities of trace gases
- Comprehensive scientific study of the Karakorum-Kunlun Mountain region
- Study of the comprehensive regional development and land restoration in Xinjiang Uighur Autonomous Region
- Study of the environmental impacts of the Three Gorges Project and possible countermeasures
- Study of the geochemistry of the boundary layer between sediment and water in lakes
- Structural composition and evolution of lithosphere in southeast China and adjacent seas
- Study of the numerical forecasting of marine environment
- Physical model of rainfall penetration, water transport in aeration zone, and the effects on bio-environment
- Pilot project demonstrating the influence of environmental and geological factors in land development in the regions around the Bohai Sea
- Study of Quaternary paleo-ocean of the South China Sea
- Sino–Japanese cooperative program on the atmosphere–land surface processes experiment (HEIFE) in the Heihe River region in Gansu Province (1987–1991) and the Sino-Japanese Cooperative Program (1990–1994)
- Generating mechanisms of red tide along China's southeast coast (1989–1992)
- Development of ecological models of grasslands in north China (1989–1992)
- Study of water-saving agriculture on the North China Plain (1992–)

- Measurement and prediction of changes and trends in the life-supporting environment in the next 20 to 50 years in China (1992–)
- Dynamic processes and environmental changes since 150,000 years before present along a corridor from Xinjiang to the Yellow Sea Continental Shelf (1992–)
- Changes in atmospheric ozone and its influence on global environment (1992–)

## Chinese National Climate Committee

### *Organization and Membership*

In 1987, the SSTC established the Chinese National Climate Committee (CNCC) (Figure 2-3). The committee's secretariat, directed by Wang Yuanzhong, is administered by SMA. Currently, the committee has 42 members:

Zou Jingmeng (chairman), SMA
Cao Jiping, SOA
Cao Pifu, SOA
Cen Jiafa, MOGM
Chen Guofan, SMA
*Chen Panqin (secretary general), Bureau of Resources and Environmental Sciences, CAS
Chou Jifan, SEDC
*Deng Nan (vice chairman), SSTC
*Fang Weiqing, Department of Science and Technology for Social Development, SSTC
Fang Zongyi, SMA
Feng Sijian, SSTC
Fu Baopu, SEDC
Gao Youxi, SEDC
Gan Zijun, CAS
Guo Dexi, SOA
*Hu Dunxin, Qingdao Institute of Oceanology, CAS
Jiang Youxu, Ministry of Forestry (MOF)
Li Zechun, SMA
Lin Jisheng, State Planning Commission (SPC)
Liu Chunzheng, MOWR
Liu Yubin (deputy secretary), SMA
Lu Jiuyuan (vice chairman), MOWR

---

*also member of the CNCIGBP.

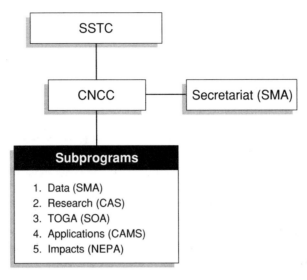

FIGURE 2-3 Organization of the Chinese national climate research program (CNCC: Chinese National Climate Committee; SMA: State Meteorological Administration; SSTC: State Science and Technology Commission).

Qiu Guangwen, Climate Bureau, Headquarters of the General Staff [military]
Shen Guoquan, SMA
Shen Wenxiong, NSFC
Tang Maocang, CAS
Tao Shiyan, Institute of Atmospheric Physics, CAS
Wang Juemou, MOWR
Wang Shaowu (vice chairman), SEDC
Wang Yangzu (vice chairman), National Environmental Protection Agency (NEPA)
Wang Yuanzhong, SMA
Weng Duming, SMA
Wu Baozhong, NEPA
Yang Wenhe (vice chairman), SOA
*Ye Duzheng (vice chairman), Institute of Atmospheric Physics, CAS
Yu Zhouwen, SOA
Zeng Qingcun (vice chairman), Institute of Atmospheric Physics, CAS
Zhang Jiacheng, SMA
Zhang Jijia (secretary, vice chairman), SMA

Zhang Qiaoling, MOA
Zhou Xiuji, SMA

The CNCC determines project activities based on proposals submitted by researchers, which are then submitted for funding through a variety of organizations, for example, CAS, NSFC, or SSTC. Mechanisms for funding are as varied as the proposals, and depend on the selection criteria of funding organizations.

A panel of experts under the CNCC has produced a compilation of Chinese climate research in a report entitled *Climate* (SSTC 1990), which is available in an English-language abridged version entitled *Climate (abridged)*. The publication provides chapters about China's climate, climate resources, major climate problems affecting economic development, past climate and possible climate changes up to the year 2050, and recommendations for response strategies.

The overall Chinese climate research program is characterized as "based on observational facts obtained by modern means and dynamic numerical simulation approach . . . with a view to predicting its variation." It is reported that

> China possesses abundant climate data in the historical records of its ancient observatories and in other kinds of writings that can serve as the valuable means of studying its long-term evolution. . . . [It] is still backward in modern observational facilities, lacks high-speed communication equipment and powerful huge-sized computers, and there is chaos in the management of climatic data kept by different departments (SSTC 1990).

The Chinese acknowledge that these institutional inefficiencies and limited infrastructure impede the progress of climate research.

### *CNCC Research Agenda*

A panel of experts under the CNCC has also produced the "Outline of the National Climate Program of China (1991-2000)" (CNCC 1990) to describe the overall nature and objectives of climate research in China. The outline presents the Chinese national climate program, which consists of five subprograms parallelling the WMO climate programs:

*Climate data.* This subprogram is located at the SMA National Meteorological Center and is concerned with collecting compatible national data sets and improving monitoring. The CNCC's outline clearly identifies the problems with data quality, lack of compatibility, disorganized and duplicating data collection and monitoring activities among separate administrative systems, and the need for a coordinated national program for climate data. The World Data Center (WDC)-D[4] for atmosphere is located in the Information Office of China's National Meteorological Center at SMA.[5]

*Climate Research.* This subprogram is located at the CAS Institute of Atmospheric Physics and is concerned with modeling and numerical simulation, and observational programs. The research subprogram at the CAS Institute of Atmospheric Physics includes WCRP radiation projects involving (a) a baseline surface radiation network and (b) a western North Pacific cloud radiation experiment, a WCRP modeling experiment on climate change involving (a) a 2-level general circulation model (GCM), (b) a 9-level GCM, and (c) a 4-level ocean GCM model, and numerical simulation projects. The Sino–Japanese HEIFE collaboration (Chapter 4) is considered a Global Energy and Water Experiment (GEWEX) under WCRP, which has been combined recently with the IGBP Biospheric Aspects of the Hydrological Cycle (BAHC) Core Project. The leading institute for the HEIFE experiment is the CAS Lanzhou Institute of Plateau Atmospheric Physics.

*Tropical Oceans Global Atmosphere (TOGA).* This subprogram (Chapter 4) is located at SOA and is concerned with data and modeling describing the coupling between ocean and atmosphere in the tropics. SOA's work includes (a) observation systems, (b) TOGA-Monsoon climate research, and (c) monitoring of *El Niño* events.

*Climate Application.* This subprogram is located at CAMS and is concerned with the use of climate resources.

*Climate Impact.* This subprogram is located at the NEPA Chinese Research Academy of Environmental Sciences (CRAES) and is concerned with the effects of climate variation and change. Specifically, CRAES will be studying trends and impacts of past climate and greenhouse gases.

## National Climate Change Coordination Group

The National Climate Change Coordination Group (NCCCG) was established in 1990 by the State Environmental Protection Commission (SEPC) of the State Council and is responsible for policy on climate change issues and interagency coordination. The group meets only as needed, usually not more than two to three times a year. Under the chairmanship and administration of SSTC (see Chapter 3), the group has four vice chairmen (the SMA administrator, NEPA administrator, SSTC vice chairman, and Ministry of Foreign Affairs [MOFA] vice minister) and has membership representing 16 organizations, for example, SSTC, NEPA, SEDC, CAS, MOFA, and SPC. Administration is through the SSTC Department of Science and Technology for Social

Development, and the group's secretariat is at SMA, under the directorship of Liu Jibin, deputy director of SMA.

The NCCCG has four working groups that parallel the ones of the Intergovernmental Panel on Climate Change (IPCC): (1) scientific assessment (lead agency is SMA), (2) impacts assessment (NEPA), (3) response strategies (SSTC), and (4) international climate change agreement negotiations (MOFA). The working group on climate impacts published a paper, "Impact of Human Activities on Climate in China" (NCCCG 1990) for the Third Plenary Conference of the IPCC Second Working Group. The paper covers impacts on agriculture, forestry, water resources, and energy.

## NOTES

1. The panel does report on the newly established international program on the human dimensions of global change in Chapter 4.

2. Activities related to regional research centers are reported under the System for Analysis, Research, and Training section in Chapter 4.

3. The ICSU Panel on World Data Centers (WDC) manages about 40 centers that are distributed among five host countries that maintain them: WDC-A in the United States, WDC-B in the former Soviet Union, WDC-C1 in Europe, WDC-C2 in Japan, and WDC-D in China. WDCs collect, distribute, and archive data that provide baseline information for disciplinary research and for monitoring changes in the geosphere or biosphere (ICSU 1992).

4. Even though WDC-D for atmosphere is located at SMA, CAS is the secretariat for WDC-D activities. It should be noted that Chinese scientists not affiliated with SMA have reported being charged for WDC-D atmospheric data handled by SMA.

5. Some of these projects are detailed in later chapters and appendices.

# 3

# Overview of Institutions Revelant to Global Change Research

## ORGANIZATION

The current organization of science in China has implications for Chinese global change research and for individuals or foreign institutions that may wish to collaborate. This chapter reports on the basic organization of major institutions that conduct research, fund, or set policies that bear significantly on the conduct of global change science in China.[1] Chinese institutions conducting global change research are organized into several separate, vertically integrated administrative systems (Figure 3-1). For example, the State Meteorological Administration (SMA) has its own Chinese Academy of Meteorological Sciences (CAMS), the Chinese Academy of Sciences (CAS) has 123 institutes covering the range of scientific disciplines, the State Oceanographic Administration (SOA) has its own research institutes, and the National Environmental Protection Agency (NEPA), an independent agency only since 1988, has already established five institutes and its own Chinese Research Academy of Environmental Sciences (CRAES). Administrative, funding, and personnel policies in turn reinforce this type of vertical organization.

The lack of internal and external disciplinary or programmatic integration of scientific institutions can lead to unnecessary duplication of efforts, and problems communicating data across institutional structures limits the effectiveness of research. The problem of paral-

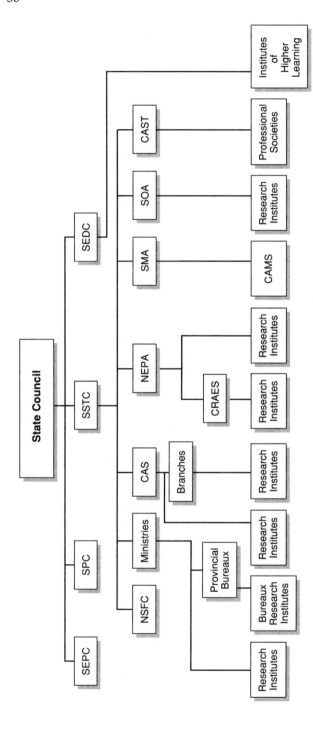

FIGURE 3-1 Simplified organization of institutions involved in Chinese global change research, policy, or funding (CAMS: Chinese Academy of Meteorological Sciences; CAS: Chinese Academy of Sciences; CAST: China Association of Science and Technology; CRAES: Chinese Research Academy of Environmental Sciences; NEPA: National Environmental Protection Agency; NSFC: National Natural Science Foundation of China; SEDC: State Education Commission; SEPC: State Environmental Protection Commission; SMA: State Meteorological Administration; State Oceanographic Administration; SPC: State Planning Commission; SSTC: State Science and Technology Commission).

lel vertical organization and resulting lack of integration is especially relevant for global change research, given the need for multi- and interdisciplinary research programs. For example, both SMA and CAS are trying to develop general circulation models—a huge and duplicative investment of scarce science resources. And, the CAS Institute of Atmospheric Physics, instead of working with the CAS Institute of Botany (a center of excellence in vegetation modeling), is contemplating starting its own ecological division for climate–vegetation interaction studies.

Chinese scientists are aware of these inefficiencies and compensate through ingenuity and individual collaborations. Fortunately, as Chinese participation in international global change research programs increases, it appears to promote larger scale multidisciplinary and interinstitutional research projects that span various administrative systems. While integration is still limited, it is a significant first step in mitigating constraints imposed by vertical organization.

One of the biggest problems resulting from the way science is organized in China is the effect on the collection, management, and accessibility of data. Emissions data collected by ministries are controlled by bureaucrats who are suspicious of any access or analysis because these data are often considered important to national security. Data collected by individual scientists often remain undocumented, and basic scientific data management principles do not appear to have been widely adopted by the research community. Data generated may often become a commodity, if they are available at all to outside parties. The problem of charging for data exists at all levels in China, a fact further complicated by the lack of interinstitutional cooperation. Often, the response to the difficulty in accessing data has been for institutions to devote their own resources to collecting their own, duplicative data.

## FUNDING

Economic reforms and tighter budgets have changed the way science is funded in China in the past decade. More and more, funding is being awarded on a grant basis for specific projects, and research institutions are responsible for raising more and more of their own funding for direct program costs. One result of these changes has been ever increasing pressure on Chinese institutions to seek international cooperation to carry out research projects, to gain access to expertise, training opportunities, and equipment.

Funding for scientific research comes from various governmental sources and reflects the organization of science by administrative systems, each of which receives its own budget for operations such as research. Consequently, "the distribution of research expenditures, the decision as to projects to proceed with and the organizing of scientific research are basically done by administrative means" (Li 1990). This organization complicates funding for global change research, which must cut across these separate administrative systems in many cases. However, the National Natural Science Foundation of China (NSFC) (see below), which was established directly under the State Council in 1986, does offer an important alternative that stresses science-driven research and funding for basic and applied basic research.[2]

Overall, annual government spending on science and technology is about 1 percent of gross national product. Annual expenditures for basic research[3] [in 1990] was "about 800 million *yuan* (approximately $200 million), accounting for 7 percent of all government investment in science and technology" (Li 1990).

For the funding of global change research, the most important institutions are probably NSFC and the State Science and Technology Administration (SSTC), followed by CAS. NSFC, as discussed in detail below, has the mandate and mechanisms for funding initiatives in global change. The biggest problem, though, are the limits of NSFC's relatively small overall budget. SSTC (see below), as a broker of funding and a coordinator for civilian science initiatives, also can play a substantive role in organizing the necessarily integrated global change research activities at the requisite scales. CAS is important, too, in that, as a separate administrative system, it could choose to make global change research a funding priority in its budget, and, given CAS' preeminent role in global change research, it would have direct and significant ramifications on China's national global change research agenda.

## CHINESE ACADEMY OF SCIENCES

The Chinese Academy of Sciences (CAS) has taken the lead in global change research for three reasons. First, the Chinese global change effort is due, in large part, to the commitment and unflagging efforts of Ye Duzheng, director *emeritus* of the Institute of Atmospheric Physics, CAS special advisor (a senior position of considerable stature), and chairman of the China National Committee for the International Geosphere-Biosphere Program (CNCIGBP). He has believed from the very beginnings of international discourse that China should be concerned about global change and should organize a scientific response. Second, CAS has a national research infrastructure covering all of the scientific disciplines relevant to global change. And third, CAS is perceived by other Chinese institutions as having responsibility for basic research, and global change is perceived as basic science.

CAS is based on the Soviet model of creating institutes organized around specific disciplines, and it currently is composed of 123 institutes, the University of Science and Technology of China in Hefei, 22 open laboratories, a few of which are also national key laboratories,[4] affiliated corporations, and a library. CAS institutes can confer degrees; 118 award master's degrees and 83 award Ph.Ds. Personnel number over 80,000, approximately two-thirds of which are scientists or technicians. In China, CAS is considered the leading scientific institution for basic research in China. Currently, approximately one-quarter of its research is considered basic, which is still more than other institutions in China (CAS undated a). However, since the early 1980s, CAS has been increasingly emphasizing applied research.

Five of nine subcenters for World Data Center (WDC)-D are located in CAS institutions, where the following databases are maintained: renewable natural resources and the environment data at the Commission for Integrated Survey of Natural Resources (CISNAR); astronomy data at the Beijing Astronomical Observatory; glaciology and geocryology data at the Lanzhou Institute of Glaciology and Geocryology; geophysical data at the Institute of Geophysics; and space science data at the Research Center for Space Science and Applications.

Institutes within CAS do collaborate, but often it is along disciplinary lines for specific larger projects, often with little or no integrative function incorporated into the project design. CAS has recently begun to reform its personnel policies to encourage movement of scientists among institutions in ways that would make collaboration easier. Also, as mentioned above, the development of the open laboratory system is specifically designed to increase staff mobility, improve the quality of research, reduce duplication, and build links to non-CAS and foreign researchers and institutions; these are important and positive steps.

Although CAS was not completely disabled during the Cultural Revolution, it was decentralized. Since the mid-1970s, CAS has recentralized somewhat, although it retains a system of 12 branches that have a voice in their local institutes' research and funding. Overall, institutes have quite a bit of autonomy, which they exercise as funding permits (Saich 1989). Further information about CAS institutes and research can be found in other chapters and in Appendix A.

## NATIONAL ENVIRONMENTAL PROTECTION AGENCY

The National Environmental Protection Agency (NEPA) began as the implementing arm of the State Environmental Protection Commission (SEPC) (see below). In 1988, NEPA's status was elevated to one of an independent agency operating under the State Council. Qu Geping, vice chairman of SEPC, is the NEPA administrator. In its short history, NEPA has made great strides in devising standards for environmental quality and establishing a discharge permit system in 230 cities, an ambient monitoring system, and an enforcement policy. NEPA's primary emphasis has been on local and regional environmental issues. It is clear that NEPA is addressing global environmental change in its work in addition to domestic efforts to address environmental issues, yet NEPA has not responded to the open invitation to join CNCIGBP.

In the fall of 1991, a United Nations Development Program (UNDP) delegation visited China to assist in the development of Chinese policies concerning the Montreal Protocol. In fact, NEPA is the agency coordinating various ministries in China to gather information for a UNDP report on recommendations for global ozone protection. Another successful effort has been collaborative work between the NEPA and the U.S. Environmental Protection Agency (EPA) on coal-bed methane, which was the basis for a $10 million grant from the Global Environment Facility[5] in 1992.

Moreover, China has received $2 million from the Global Environment Facility to conduct a 2-year study on the control and mitigation of greenhouse gases that will be implemented through NEPA. This study will include a study of potential impacts of climate change during the next 50 to 100 years by reviewing models for sea-level rise and other impacts of climate change to assess human, social, and economic impacts, by assessing various mitigation strategies to reduce greenhouse gas emissions, and by making policy recommendations.

NEPA is conducting studies of existing and projected chlorofluorocarbon (CFC) production and technology needs in order to come up with a plan for phasing out CFCs over the next 10 to 20 years, if not sooner. NEPA and EPA have organized technical assistance projects for the Chinese to learn more about alternative CFC technologies. For example, China is developing alternatives to CFC refrigeration. NEPA is especially interested in conservation technologies (for example, recycling and reclamation) as a means of meeting the growing demand for CFCs in China. China estimates that its 1996 CFC production will be 60,000 tons, but the Ministry of Chemical Industry has stepped up efforts to find alternatives since China became a Montreal Protocol signatory.

In 1991, construction began on the Sino–Japanese Friendship Environmental Protection Center at NEPA in Beijing. Total cost of the center is estimated to be approximately $70 million. When completed in 1994, the center will house six departments: (1) environmental monitoring, (2) environmental pollution control technology, (3) environmental information, (4) environmental strategy and policy, (5) environmental training and education, and (6) administration. Staff size is projected to be 500. It is expected that this center will develop programs relating to global environment; however, it is too early to know more with great certainty or detail. NEPA plans to have a national monitoring system and a national environmental information system administered through this center.

NEPA maintains its own research academy, CRAES, in Beijing, the Nanjing Institute of Environmental Science (Appendix A), the South China Institute of Environmental Protection, the Xinjiang Institute of Environmental Research, the Institute of Environmental and Economic Policy Research, and the Wuhan University Institute of Environmental Law Research.

Under CRAES, which was founded in 1979, various institutes conduct research in most aspects of environmental studies: the Institute of Atmospheric Environment, Institute of Water Environment, Institute of Ecological Environment, Institute of Analysis and Measurement, Institute of Environmental Management, Institute of Environmental Standards, Institute of Environmental Information, and the Center of Computation (which has environmental databases). Research projects are usually funded by NEPA and SSTC.

Since 1988, CRAES has sponsored the journal, *Environmental Science Research* (*Huanjing Kexue Yanjiu*), the chief editor of which is Liu Hongliang. CRAES staff also edit the journals *Environmental Science and Technology Information* (*Huanjing Keji Qingbao*) and *Translation of Environmental Science* (*Huanjing Kexue Yicong*). Also, Wang Wenxing, a professor of environmental chemistry at CRAES, is the chief editor of *China Environmental Science* (*Zhongguo Huanjing Kexue*), a leading journal that is published bimonthly in Chinese.

## NATIONAL NATURAL SCIENCE FOUNDATION OF CHINA

The National Natural Science Foundation of China (NSFC) was established in 1986 under the State Council, although it functions independently. Prompted in part by the establishment of the International Geosphere and Biosphere Program (IGBP), NSFC has directed its attention to the environmental and, in particular, global change sciences. The NSFC's *Guide to Programs* (NSFC 1990) indicates a high degree of sophistication in its understanding of global change as an emerging scientific discipline, as well as of its implications for economic development. It is eager to support and to participate in IGBP and World Climate Research Program (WCRP) strategies in the study of global change, and to coordinate its research effort in jointly funded national and international projects.

The main task of the NSFC is to "guide, coordinate, and support[6] basic research and part of applied research" (Li 1990). Unlike other funding sources, applications to NSFC are subjected to a peer review process by scientific experts. Furthermore, "scientists are responsible to the NSFC only, and they may independently decide how to use the grants provided, and how to implement the projects" (Li 1990). NSFC supports studies of the impact of human activities at the interfaces between layers: lithosphere, hydrosphere, cryosphere, atmosphere, and biosphere. Still, the NSFC recognizes that, with China's weak economic, scientific, and technological base, as well as its limited financial resources, its smaller scale projects will be limited for the foreseeable future to those fundamental issues that reflect the country's own interests, for example, those concerned with economic development, and that take advantage of the country's expertise.

NSFC projects are divided into three categories: general, key, and major. General projects receive relatively small grants, and awards deliberately support the widest range possible of ideas, people, and geographical location. General projects usually receive between 60,000 ($11,000) and 70,000 ($12,700) *yuan* over 3 years. Key projects are selected from the general pool of applications for their academic significance and for their potential applications. Key project funding usually is between 500,000 ($91,000) and 700,000 ($127,000) *yuan* for 3 years. Major projects are comprehensive topics that are selected for their importance to the development of science and to Chinese economic development goals. Funding levels begin at 2 million *yuan* ($364,000) for 5 years (Li 1990). Appendix B provides an extensive listing of general, key, and major projects relevant to global change. In general, NSFC plans to emphasize paleoclimatic study (improved sensitivity), near-term climate change, and biogeochemical interaction mechanisms.

The NSFC is divided into six departments and one group: (1) Department of Mathematical and Physical Sciences, (2) Department of Chemistry and Chemical Engineering, (3) Department of Life Sciences, (4) Department of Earth Sciences, (5) Department of Material

TABLE 3-1 Global change project funding from the National Natural Science Foundation of China, 1986-1992 ($1=5.5 *yuan*)

|  | Number of Projects | Cost |
|---|---|---|
| Major Projects | 9 | $ 4,545,455 |
| Key Projects | 20 | 181,818 |
| General Projects | 70 | 545,455 |
| Total | 99 | 5,272,727 |

and Engineering Sciences, (6) the Department of Information Science, and (7) the Management Science Group.

Between 1986 and 1992, NSFC reported funding 99 global change projects (Table 3-1). Six global change disciplines are handled by the Department of Earth Sciences: (1) geography (including remote sensing) and soil science, (2) geological sciences, (3) geochemistry, (4) geophysics and space physics, (5) atmospheric sciences, and (6) ocean sciences (Table 3-2).

The total 1990 program budget for NSFC was 160 million *yuan* ($29,090,000),[6] marking the fifth consecutive year of increased funding. The projected 1990 budget for the Department of Earth Sciences was more than 14 million *yuan* ($2,576,364). Out of 1,659 applications, 371 projects (22.4 percent) received awards. The average support per project was approximately 48,500 *yuan* ($8,800). See Appendix B for a recent listing of global change projects that have received NSFC funding.

## STATE EDUCATION COMMISSION

The State Education Commission (SEDC), directed by Li Tieying, was established by the State Council in 1985 to reform, manage, and further develop the educational system that had previously been under the aegis of the Ministry of Education. Although it has ultimate responsibility for colleges and universities, direct supervisory control is often delegated to provincial education commissions, ministries or other relevant cosponsoring institutions, for example, CAS (Reed 1989).

Chinese institutions of higher education have designations and are organized differently than in the West. National "key" institutions of higher education are the most prestigious, receive more funding, attract the best students, and have the best facilities. Provincial "key"

TABLE 3-2 Project awards in the Department of Earth Sciences, National Natural Science Foundation of China, 1987-1989 ($1=5.5 *yuan*)

| Name of Discipline | 1987 Projects | 1987 Amount | 1988 Projects | 1988 Amount | 1989 Projects | 1989 Amount | Projected Funding for 1990 |
|---|---|---|---|---|---|---|---|
| Geography and Soil Sciences | 74 | $ 485,455 | 63 | $ 523,636 | 69 | $ 598,182 | $ 612,727 |
| Geological Sciences | 111 | 883,636 | 79 | 803,636 | 95 | 994,545 | 940,000 |
| Geochemistry | 21 | 160,000 | 26 | 169,091 | 32 | 272,727 | 241,818 |
| Geophysics and Space Physics | 43 | 290,727 | 33 | 303,636 | 40 | 370,909 | 345,455 |
| Atmospheric Sciences | 22 | 132,727 | 24 | 163,636 | 26 | 184,545 | 189,091 |
| Ocean Sciences | 28 | 158,182 | 22 | 200,000 | 23 | 227,273 | 247,273 |
| Total | 299 | 2,110,727 | 247 | 2,163,636 | 285 | 2,648,182 | 2,576,364 |

institutions rank just behind the ones at the national level. The Chinese adopted the Soviet-style of educational organization in the 1950s, which organized institutions according to narrow specialties (Reed 1989). As the report has noted in reference to the organization of research institutions, organizing along narrow specializations limits opportunities for inter- and multidisciplinary collaborations.

Of the educational institutions relevant to this report, SEDC administers a range of institutions of higher learning. Comprehensive universities, for example Peking, Fudan, Nanjing, and Zhongshan, offer basic and applied sciences and social sciences and/or humanities. Polytechnical universities, for example, Qinghua, Tongji, Zhejiang, and Shanghai Jiaotong, offer applied sciences and engineering. Institutions of science and technology offer specialized training in specific areas of basic sciences or engineering, and they are often administered in conjunction with respective ministries. "Normal" colleges and universities train teachers, and Beijing Normal University and the East China Normal University are the most prestigious examples. Agricultural universities and colleges are administered in conjunction with the Ministry of Agriculture (MOA) or its bureaux (Reed 1989).

SEDC is not really involved in the graduate programs in CAS research institutes, although CAS does use SEDC entrance examinations in selecting students. SEDC does administer with CAS the CAS University of Science and Technology of China in Hefei and the CAS Graduate School for Science and Technology in Beijing.

The panel found that SEDC has not responded to the global change agenda in any formal way at the national level, although individual universities administered by SEDC are involved in research related to global change. In the cases where the Chinese noted university activities as part of the global change program, the panel requested further information and visited those institutions where possible. Further information about those projects is found in the next chapters and appendixes of this report.

Although a more detailed examination of current educational enrollment and curricula was beyond the scope and resources of this study, the panel did identify two innovative initiatives in global change education that signal interest and instances of creative response on the part of the Chinese: CAS is developing a proposal for the advanced training of scientific personnel and Beijing Normal University is developing a proposal for the basic training of pre-college teachers. The two new Chinese education initiatives focus on global environmental change at the largest scales, which the planners recognize requires international understanding to manage effectively.

The CAS initiative is particularly strong because it promotes global change education expressly for developing countries in Asia. Advanced training would be provided during an intensive 3-week summer institute for postgraduate students, recent Ph.Ds, and younger scientists currently studying or conducting research in IGBP-related areas. About one-third of the participants would be chosen from other Asian countries. This proposal has been endorsed in principle by the scientific committee of the IGBP. The leading organizer of this proposal is Fu Congbin at the Institute of Atmospheric Physics.

Beijing Normal University is proposing a summer school for advanced teacher training in global change science, which will target current and prospective teachers of pre-college and beginning higher general education students. This university has offered intensive summer environmental science courses in previous years, and the proposed new initiative builds on that experience. Planning is now under way by Wang Huadong, director of the Beijing Normal's Institute of Environmental Sciences and Xu Jialin. (Wang and Xu have authored Chinese language textbooks of environmental sciences and are also coauthors of *The Natural History of China*.) In addition, active participation is expected by Liu Jingyi, past director of the CAS Research Center for Eco-Environmental Sciences, and by the Chinese Society of Environmental Sciences, in which all three of these scientists are active.

This proposed effort to conduct innovative environmental education in China would introduce new subject matter about global change issues and also provide an opportunity for new methods of presenting complex workings of environmental systems. The participants are expected to develop new skills to help them become innovators in future environmental management or in education.

The panel also identified a tripartite agreement signed among Peking University's Center for Environmental Sciences, the Russian Academy of Sciences, and the University of Michigan in 1991 to hold annual workshops to promote international interdisciplinary research and training on global change. Tang Xiaoyan is the principal Chinese investigator and Thomas Donahue is the principal American investigator. The first meeting was supposed to be held in the former Soviet Union in 1992 but it was canceled. While details about training activities were not identified, this endeavor appears to have interesting potential.

The requirements of global change research in any country are creating pressures on educational systems that are not currently designed to respond very well to these demands. It is welcome to see that the importance of global change education is prompting proposed programs in China, as it should elsewhere without further delay.

## STATE ENVIRONMENTAL PROTECTION COMMISSION

The State Environmental Protection Commission (SEPC) was established by the State Council in 1984 and is the leading decision-making body in China for environmental issues. Membership includes the heads of all relevant ministries and agencies (SSTC, State Economic Commission, State Planning Commission, Ministry of Urban and Rural Construction, Ministry of Forestry, MOA, Ministry of Water Resources, various industrial ministries, and the Ministry of Energy), who meet quarterly to review and set environmental policy and provide interagency coordination. NEPA is the SEPC secretariat.

## STATE METEOROLOGICAL ADMINISTRATION

The State Meteorological Administration (SMA), under the direction of Zou Jingmeng, has three major centers, the National Meteorological Center,[7] the National Satellite Meteorological Center, and CAMS. Additionally, the administrative offices for the Chinese National Climate Committee and the National Climate Change Coordination Group (NCCCG) are located at SMA (Chapter 2). Of these major centers, CAMS is where SMA's research is conducted. Further details about CAMS and global change research being conducted there can be found in Chapters 4 and 5 and Appendix A.

Meteorological data in China are particularly robust and have a historical depth unmatched by any other country. Currently, SMA has more than 2,600 surface observational stations, upper-air sounding stations, weather radar stations, and various specialized meteorological stations. The monitoring network has six regional centers (Shanghai, Wuhan, Shenyang, Guangzhou, Chengdu, and Lanzhou) that are operational or under construction. Each province has a meteorological bureau, each prefecture has a weather office, and each county has a weather station. At the national level, the National Meteorological Center provides data and information, climate analysis, and other meteorological services. SMA has two satellites transmitting Advanced Very High Resolution Radiometer (AVHRR) data. These data are handled by SMA's National Meteorological Satellite Center.

## STATE OCEANOGRAPHIC ADMINISTRATION

The State Oceanographic Administration (SOA), founded in 1964 and directed by Yan Hongmo, is an active participant in Tropical

Ocean and Global Atmosphere (TOGA) (approximately 100 scientists have been involved) and World Ocean Circulation Experiment (WOCE) research. Additionally, SOA has collaborated on a joint air–sea interaction project with U.S. researchers and a cooperative project with the Japanese to study the Kuroshio Ocean Current. Further details can be found in Chapters 4 and 5.

SOA is responsible for ocean surveys, oceanographic research and monitoring, and marine resource management. SOA has three branch offices: Qingdao for activities related to the North Sea, Shanghai for activities related to the East China Sea, and Guangzhou for activities related to the South China Sea. SOA operates the First, Second, and Third Institutes of Oceanography, the Institute of Marine Environmental Protection, the Institute of Ocean Technology, the Institute of Marine Scientific and Technological Information,[8] the Institute of Desalination, the Marine Environmental Forecasting Center, the China Ocean Press, and the Ningbo Oceanography School. It operates 42 research and monitoring vessels (four are 10 kilotons or heavier), two remote sensing aircraft, eight central marine stations, 58 marine investigation stations, three monitoring and surveillance centers, 12 data buoys, and three regional forecasting stations.

## STATE PLANNING COMMISSION

The State Planning Commission (SPC), under the chairmanship of Vice Premier Zou Jiahua is the largest comprehensive department under the State Council and the highest administrative organ for macroeconomic management. In China's centrally planned economy, SPC is responsible for producing the 5-year plans, which detail the overall budgets for institutions such as CAS, NSFC, SSTC, and the ministries. SPC awards key project grants that provide fairly substantial funding. In general, the trend in government has been away from block financing to institutions in favor of project specific grants.

According to Li Fuxian, deputy director of the SPC National and Regional Planning and Development Bureau, SPC, like other Chinese agencies, is committed to working with SSTC, NEPA, SMA, and other relevant ministries to study global environmental issues. To that end, SPC was charged with coordinating the production of the "National Report of the People's Republic of China on Environment and Development" for the United Nations Conference on Environment and Development (UNCED), which had a specific section on the impact of global warming on China (SPC 1991).

## STATE SCIENCE AND TECHNOLOGY COMMISSION

The State Science and Technology Commission (SSTC), which operates directly under the State Council, is the leading administrative agency for civilian science and technology. Song Jian is the chairman and he also is a state counselor, chairman of SEPC, and chairman of various other environmental committees, including being the honorary chairman of the CNCIGBP.

SSTC is responsible for developing science and technology policy and guidelines and for applying science and technology for economic growth. Budgets for research and development in science and technology are administered through SSTC, including the CAS and other research institutes in various government agencies. Clearly, the strength of Chinese global change research is dependent upon favorable support by SSTC.

Of the 100 million *yuan* budgeted for environmental protection research in the Eighth 5-Year Plan, SSTC is allocating 80 million (NEPA is contributing the remainder). Six major environment projects, which concentrate on atmosphere and water pollution, have been identified, and responses to global climate change is one.

In 1989, the Department of Science and Technology for Social Development was established and was originally headed by Deng Xiaoping's daughter, Deng Nan, (vice chairman of the CNCIGBP). This department, now under the direction of Gan Shijun, is responsible for coordinating funding and overall administration of many projects, including those related to environment and global change. Within a given project, detailed project management is delegated to groups of experts.

The department is a member of NCCCG (Chapter 2). As a member, SSTC chairs the working group for response strategies, and, to this end, will develop a national response strategy for global climate change with financing from the Asian Development Bank. SSTC will assess current and projected greenhouse gas emissions, review strategies for reducing them, and analyze policy implications. Additionally, SSTC selected the study of changes in the life-supporting environment in the next 20 to 50 years to be one of the national key projects under the Eighth 5-Year Plan, for which Ye Duzheng, CAS, is the lead scientist (Chapter 2). A large-scale, multidisciplinary project, The Origin, Evolution, Environmental Changes, and Ecosystems of the Qinghai–Tibet Plateau is administered by SSTC (Chapter 4). The SSTC project coordinator for global climate change projects is Wen Jianping.

SSTC administers the National Remote Sensing Center, but actual research and training are carried out at various research institutions.

## NOTES

1. The panel decided that, although some ministries do control important emissions data that are relevant to global change research, a discussion of their overall roles does not fit the objectives of this chapter. Relevant research activities of various ministries are reported elsewhere in this report.

2. Applied basic research is defined here as basic research that is oriented to some application (Li 1990).

3. Basic research is defined here as "mainly aimed at understanding natural phenomena, discovering objective laws and conducting systematic investigation, examination, and analysis of basic scientific data and exploring the associated laws" (Li 1990).

4. Open laboratories are a CAS designation that means the facility has met certain standards for operation and equipment and that it is open to non-CAS organizations and foreign scientists. These laboratories usually receive 100 to 200 percent higher funding from CAS. Staff mobility, openness to outside researchers, graduate training, multi-institutional and international collaboration are stressed. National key laboratories are similar to, and took their lead from, CAS' open laboratories. In fact, all national key laboratories at CAS are also open laboratories. National key laboratories at CAS stress applied over basic research. While CAS has developed and runs open laboratories, most of the national key laboratories are run by the State Education Commission (SEDC). The basic difference between them is that open laboratories stress research and national key laboratories stress education (Abramson 1990).

5. The Global Environment Facility is a multilateral fund set up by governments, the World Bank, the United Nations Environment Program, and the United Nations Development Program to finance grants and low-interest loans to developing countries for projects related to global environment, for example, greenhouse gas response strategies, biodiversity action plans, and technology transfers.

6. NSFC funding covers only direct project costs.

7. The WDC-D subcenter for atmospheric data is located in the Information Office of this center.

8. The WDC-D subcenter for oceanography data is located in this institute.

# 4

# Chinese Participation in International Global Change Research Programs

### INTRODUCTION

This chapter presents an overview of Chinese activities in three major international global change programs: (1) the International Geosphere–Biosphere Program (IGBP), which is sponsored by the International Council of Scientific Unions (ICSU), (2) the World Climate Research Program (WCRP), which is sponsored jointly by the World Meteorological Organization and ICSU, and (3) the Human Dimensions of Global Environmental Change (HD/GEC) Program, which is sponsored by the International Social Science Council.

Research highlights are presented in each of the core project areas in which China is, plans to be, or has the potential (in the panel's view) to be actively engaged. A section on the Chinese Ecological Research Network (CERN) is also included as it is a component of the CNCIGBP's global change program. Further details about the organization and research of selected institutions identified in this chapter are provided in Appendix A.

Below are listed Chinese institutes conducting research related to IGBP, WCRP, or HD/GEC that are identified in this report.

**Biospheric Aspects of the Hydrological Cycle and Global Energy and Water Cycle Experiment (BAHC/GEWEX) IGBP, WCRP**
Institute of Geography, CAS
Lanzhou Institute of Plateau Atmospheric Physics, CAS
Shanghai Institute of Plant Physiology, CAS

***Data and Information Systems (DIS), IGBP***
  Institute of Atmospheric Physics, CAS

***Global Analysis, Interpretation, and Modeling (GAIM), IGBP***
  Institute of Atmospheric Physics, CAS
  Institute of Botany, CAS
  Chinese Academy of Meteorological Sciences, SMA
  Peking University
  Shanghai Institute of Plant Physiology, CAS
  State Oceanographic Administration
  University of Science and Technology of China, CAS

***Global Change and Terrestrial Ecosystems (GCTE), IGBP***
  Institute of Botany, CAS
  Shanghai Institute of Plant Physiology, CAS

***Human Dimensions of Global Environmental Change (HD/GEC), ISSC***
  Commission for Integrated Survey of Natural Resources, CAS
  Guangzhou Institute of Geography
  Institute of Automation, CAS
  Institute of Systems Science, CAS
  Ministry of Aviation and Space Flight
  Nanjing Institute of Geography and Limnology, CAS
  Research Center for Eco-Environmental Sciences, CAS
  State Science and Technology Commission
  Xinjiang Institute of Geography, CAS

***International Global Atmospheric Chemistry Project (IGAC), IGBP***
  Anhui Institute of Optics and Fine Mechanics, CAS
  Beijing Municipal Academy of Agriculture and Forestry Services
  Chinese Academy of Meteorological Sciences, SMA
  Institute of Atmospheric Physics, CAS
  Nanjing University
  Peking University
  Research Center for Eco-Environmental Sciences, CAS
  South China Institute of Botany, CAS
  University of Science and Technology of China, CAS

***Joint Global Ocean Flux Study (JGOFS), IGBP; Land–Ocean Interactions in the Coastal Zone (LOICZ), IGBP***
  First Institute of Oceanography, SOA
  Guangzhou Institute of Geography
  Institute of Geography, CAS
  Lanzhou Institute of Desert Research, CAS

Nanjing Institute of Geography and Limnology, CAS
Nanjing University, Department of Geo and Ocean Sciences
Qingdao Institute of Oceanology, CAS
Qingdao University of Oceanology
Second Institute of Oceanography, SOA
South China Sea Institute of Oceanology, CAS
Third Institute of Oceanography, SOA
Xiamen University, Department of Oceanography

*Past Global Changes (PAGES), IGBP*
Beijing Normal University, Department of Geography
Chinese Academy of Meteorological Sciences, SMA
Chinese University of Geosciences, Wuhan
Commission for Integrated Survey of Natural Resources, CAS
Fudan University
Guiyang Institute of Geochemistry, CAS
Institute of Atmospheric Physics, CAS
Institute of Botany, CAS
Institute of Geography, CAS
Institute of Geology, CAS
Institute of Geology, CAGS
Institute of Oceanographic Geology, MOGM
Kunming Institute of Botany, CAS
Lanzhou Institute of Glaciology and Geocryology, CAS
Lanzhou Institute of Desert Research, CAS
Nanjing Institute of Geography and Limnology, CAS
Nanjing University
National Remote Sensing Center, CAS
Northeast Normal University
Peking University
Qinghai Institute of Saline Lakes, CAS
Shaanxi Normal University
Shenyang Institute of Applied Ecology, CAS
South China Sea Institute of Oceanology, CAS
Third Institute of Oceanography, SOA
Tongji University
Xi'an Laboratory of Loess and Quaternary Geology, CAS
Xinjiang University, Department of Geography
Yunnan Institute of Geological Sciences
Zhongshan University

*System for Analysis, Research, and Training (START), IGBP*
Institute of Atmospheric Physics, CAS
Institute of Botany, CAS

Research Center for Eco-Environmental Sciences, CAS
Xi'an Laboratory of Loess and Quaternary Geology, CAS

**Tropical Ocean and Global Atmosphere Program (TOGA), WCRP**
Chinese Academy of Meteorological Sciences, SMA
First Institute of Oceanography, SOA
Institute of Atmospheric Physics, CAS
Institute of Geography, CAS
Institute of Mechanics, CAS
Lanzhou Institute of Plateau Atmospheric Physics, CAS
Qingdao Institute of Oceanology, CAS
Qingdao University of Oceanology
Second Institute of Oceanography, SOA
South China Sea Institute of Oceanology, CAS
Third Institute of Oceanography, SOA

## INTERNATIONAL GLOBAL ATMOSPHERIC CHEMISTRY PROJECT

The International Global Atmospheric Chemistry Project (IGAC) was created under the auspices of the Commission on Atmospheric Chemistry and Global Pollution in 1988 in response to the growing international concern over observed changes in atmospheric chemical compositions and their potential impact on mankind. When the IGBP was formed, IGAC was adopted as one of its core projects. The overall goal of IGAC is to measure, understand, and thereby predict changes—now and over the next century—in global atmospheric chemistry, with emphasis on changes affecting the oxidizing capacity of the atmosphere, the impact of atmospheric composition on climate, and the interactions of atmospheric chemistry with the biota. The goal is broad and encompasses several contemporary environmental issues, including the increased acidity of precipitation, the depletion of stratospheric ozone ($O_3$), and global warming due to the accumulation of greenhouse gases, for example, carbon dioxide ($CO_2$), methane ($CH_4$), nitrous oxide ($N_2O$), and chlorofluorocarbons (CFCs).

According to the IGAC plan, six major foci address important problems in global atmospheric chemistry, whose solutions require international cooperation: (1) natural variability and anthropogenic perturbations of the marine atmosphere; (2) natural variability and anthropogenic perturbations of tropical atmospheric chemistry; (3) role of polar regions in changing atmospheric composition; (4) role of boreal regions in changing atmospheric composition; (5) global distribution, transformation trends, and modeling; and (6) international support activities.

## Research Highlights

Atmospheric chemistry research is carried out in a number of institutes and universities, usually addressing urban air pollution issues such as oxidants, suspended particles, and toxic species. Recently, some attention has been directed at research projects that have regional and global implications. Most of these projects are closely related to IGAC research activities. A major focus is on greenhouse gas emissions, including $CH_4$, $N_2O$, and $CO_2$. Research projects on stratospheric $O_3$ have also been carried out. Regional-scale research activities are focused on acid precipitation and oxidants. In addition, the interesting problem of long-range transport of Asian dust and its impact on the Pacific Basin has also drawn some attention. A brief description of these projects is presented here and additional details are discussed in Chapter 5.

The panel identified significant interest in studying $CH_4$ emissions in China. Observed to be increasing at the rate of about 1 percent per year, $CH_4$ is one of the important trace gases implicated in global warming. With present day concentrations of about 1.75 ppm, increases in $CH_4$ can affect global climate as well as tropospheric and stratospheric chemistry. Rice paddy fields—of which 24 percent of the world's total lie in China—are an important source of atmospheric $CH_4$, contributing to approximately 10 to 20 percent of total global emissions. At least four groups, the Chinese Academy of Sciences (CAS) Institute of Atmospheric Physics, Chinese Academy of Meteorological Sciences (CAMS), CAS Research Center for Eco-Environmental Sciences (RCEES), and CAS Nanjing Institute of Soil Science, have either made or started measurements of $CH_4$ emissions from rice fields. Some of the measurements were conducted as bilateral collaborations between the CAS Institute of Atmospheric Physics and the Fraunhofer Institute in Germany (Wang et al. 1992) and between the CAS Institute of Atmospheric Physics and the U.S Department of Energy (DOE)[1] (Khalil et al. 1990). These groups have also started measurements of $N_2O$ emissions from soils. In addition, exchange of $CO_2$ between the biosphere and the atmosphere is being studied at the CAS South China Institute of Botany and RCEES.

Trace gas and aerosol monitoring is carried out by various organizations. CAMS, RCEES, and Peking University operate several atmosphere stations in rural and remote areas where measurements of

$O_3$, sulfur dioxide ($SO_2$), aerosols, and precipitation chemistry are carried out. In addition, networks of stations have been established for national-scale precipitation chemistry measurements by NEPA and CAMS and regional-scale measurements by RCEES and several provincial agencies.

Aerosol chemistry studies also focus primarily on urban and regional-scale environmental problems. An exception is the study of Asian dust storms that exert a large influence on the chemical and physical characteristics of aerosols and precipitation over eastern Asia and the northern Pacific Ocean. CAMS has an extensive program to study the formation and transport of dust storms. A cooperative international program, the China and America Air-Sea Experiments (CHAASE), has been conducted to study the compositions of aerosol particles and precipitation in China and Korea since 1990 (Arimoto et al. 1990, Gao et al. 1992a,b). The program involves the State Oceanographic Administration (SOA), the Korean Ocean Research and Development Institute, and the University of Rhode Island.[2] In addition, as part of the WCRP Tropical Ocean and Global Atmosphere (TOGA) program (see below), compositions of rain and aerosol samples collected over the western Pacific Ocean were analyzed under bilateral projects between the U.S. National Oceanographic and Atmospheric Administration (NOAA) and the State Meteorological Administration (SMA) and between NOAA and the Chinese National Research Center for Marine Environment Forecasts at the State Oceanographic Administration (SOA).

Total $O_3$ is measured by scientists from the CAS Institute of Atmospheric Physics at a station in Beijing and one in Yunnan Province. Ground-based remote sensing techniques for measuring stratospheric trace gases such as $O_3$ and nitrite ($NO_2$) are under development at CAS Anhui Institute of Optics and Fine Mechanics and Peking University. Modeling studies of the stratospheric $O_3$ are conducted at the CAS Institute of Atmospheric Physics, SMA, CAS University of Science and Technology of China, and Peking University. In addition, chemistry models of the troposphere are also under development at some of these institutes for the study of regional and global environmental problems.

In September and October 1991, the U.S. National Aeronautics and Space Administration (NASA) conducted the first of its four planned airborne experiments over the Pacific Basin as part of what are collectively called the Pacific Exploratory Mission (PEM). Their major objectives are to investigate the budgets of tropospheric oxidants, reactive nitrogen species, and sulfur species. The first of these experiments, known as PEM–West, is coordinated through the East Asia–

North Pacific Regional Study (APARE) of the IGAC Program, which includes scientists from China, Hong Kong, Japan, Korea, Taiwan, and the United States. PEM–West scientists measured a suite of important trace gases and aerosols from two aircraft (a NASA DC-8 and a short-range Japanese aircraft)[3] and six intensive ground stations over the western Pacific Ocean. SMA, in collaboration with NOAA, operates one such intensive station on the eastern coast of China.

## PAST GLOBAL CHANGES

The objective of the IGBP Past Global Changes (PAGES) Core Project[4] is to organize efforts internationally to better understand past changes in the earth system in order to put current and future global changes into perspective and to improve the interpretation of their causes and dynamics. PAGES has taken a "two-stream" approach. The first stream is directed to relatively recent earth history of the last 2,000 years. The second stream takes a longer view of the glacial-interglacial cycles of the Late Quaternary Period (IGBP 1990).

### Review of CNCIGBP Literature

Research on historical analysis of environmental change is voluminous in China; virtually every aspect of PAGES research described in IGBP Report No. 12 (1990) or in Bradley (1991) is being reported. Every CAS institute involved in global change research lists some form of historical analysis (CAS 1991), and the National Natural Science Foundation of China (NSFC) has funded this area extensively (Appendix B). In fact, the literature is so enormous that it would require a separate and extensive inquiry to catalogue and review materials cited by the Chinese. With few exceptions, work identified by the panel was restricted to China and connections with the rest of the earth system such as telecommunications with global climate anomalies remain to be made.

The fundamental objectives of this work have been summarized by the CNCIGBP (Ma 1991):

1. Reconstruction of past climate change and environmental variation, especially covering the last 2,000 years, through the enormous Chinese historical writings on climate and environmental descriptions, especially in eastern China.

2. Development of multiproxy data from tree-ring chronologies, archaeological studies, and ice core and sedimentary analyses to supplement written records.

3. Study of glacial-interglacial cycles in the Late Quaternary Period, with special attention toward rapid and abrupt changes in order to better understand future global changes.

4. Establishment of an historical proxy data bank in China.

5. Performance of some historical analyses on particularly critical areas of China.

The January 1990 Report of the Chinese National Committee for IGBP (CNCIGBP 1990a) describes an element of a core project titled, "Studies of Historical Evolution of Environment." This is elaborated on more completely in the September 1990 report (CNCIGBP 1990b), which lists a number of objectives based mainly on the organization developed by IGBP. A focus on PAGES research is prominent, including three definite activities: (1) pilot assessment of the current state of the life-supporting environment in China; (2) study of the historical evolution of life-supporting environment in China; and (3) impacts of paleoclimate change on underground water resources in the Late Pleistocene Epoch and trends of climate change in arid and semiarid areas. The second activity, in particular, has numerous studies listed. Other historical analyses are imbedded in the GCTE section of that document.

As mentioned in the introduction of the CNCIGBP (Chapter 2), the *Bulletin of the CNCIGBP*, (CNCIGBP 1991) lists three focal areas for past global change research. Under "Ongoing research projects . . .," several might incorporate historical analyses although it is not explicitly stated. The final section of this bulletin gives an overview of pilot studies conducted between 1988 and 1991 that clearly put climate change in an historical perspective.

## Research Highlights

This pervasive consciousness of the variable past has elevated paleoenvironmental studies to a much higher relative level of priority in the Chinese global change program than in the United States or Europe. Every institute or laboratory that panel members visited included a paleoenvironmental component in its overall global change program. The range of records being studied included everything from the analysis of Chinese Imperial Court records at the CAS Institute of Geography (Appendix A) to the study of isotopic time series from ice cores from mid-latitude glaciers at the CAS Lanzhou Institute of Glaciology and Geocryology (Appendix A) and loess deposits by the CAS Xi'an Laboratory of Loess and Quaternary Geology (Appendix A). In each of these cases, American researchers are collaborators.

This research also demonstrates the unique paleoenvironmental records in China that are a crucial resource for the global PAGES community. The quality of scholarship and skill in this area seemed very high. The relatively modest cost and technological level required for many paleoenvironmental studies undoubtedly aid in the success of these programs. But this research is also well-supported with sophisticated instrumentation at the CAS Lanzhou Institute of Glaciology and Geocryology and the CAS Nanjing Institute of Geography and Limnology (Appendix A), two key institutions for research on paleoenvironments.

A major project, "The Origin, Evolution, Environmental Change, and Ecosystems of the Qinghai–Tibet Plateau," is funded by State Science and Technology Commission, with contributions from participants (20 units from CAS and universities are involved). This project is notable for its scale, multidisciplinary scope, and the cross-institutional organization of research. This project has four components: (1) structure, evolution of lithosphere and geodynamics (lead principal investigators (PIs): Pan Yushen, CAS Institute of Geology and Kong Xiangru, CAS Institute of Geophysics), (2) environmental changes during the Late Cenozoic Era (Shi Yafeng, CAS Lanzhou Institute of Glaciology and Geocryology and Li Jijun, Lanzhou University), (3) monitoring and prediction of recent climate change and its environmental impact (Tang Maocang and Cheng Guodong, CAS Lanzhou Institute of Glaciology and Geocryology), and (4) the structure, function, evolution, and differentiation of ecosystems (Zheng Du, CAS Institute of Geography and Zhang Xinshi, CAS Institute of Botany). The project is headed by Sun Honglie, CAS vice president, and is scheduled to run from 1992 to 1996. Although research design details were not available to the panel, this combination of scope and topics shows a welcome opportunity to couple the past, present, and future.

## Summary of PAGES Research

Historical analyses that could be related to the PAGES Core Project are replete in past and planned Chinese research efforts. This is an area where China already has made significant contributions to the study of paleoclimate; the physical and historical resources available for analysis are substantial. The Chinese have a strong tradition in this approach, and the Chinese global change program is building on this foundation. The major study of the Qinghai–Tibet Plateau is a good example of progress being made to expand and integrate work on paleoenvironments. Chinese efforts could be further strengthened by considering the relationship between past changes and the broader global system that may have driven some of these phenomena.

Historical analysis is voluminous, and integration with prospective research (designed to provide projections of and scenarios for future climate, land use, and other important environmental changes) will further enhance their contributions in this area.

## GLOBAL CHANGE AND TERRESTRIAL ECOSYSTEMS

Global Change and Terrestrial Ecosystems (GCTE) is an IGBP core project to develop the capability to predict the effects of changes in climate, $CO_2$ concentration, and land use on terrestrial ecosystems and how these changes can lead to feedbacks to the physical and chemical climate system.

### China's Vegetation and Climate

The vegetation zones in China range from tropical rainforest and monsoon forests in southern China to boreal forests in the north, and from east to west the vegetation grades from humid forests and intensive agriculture to temperate steppes and extreme deserts, with agriculture confined to areas where water for irrigation flows from the high mountains. The topography dominated by the Qinghai–Tibet Plateau not only provides large expanses occupied by alpine vegetation, permafrost, and ice, but the plateau itself also plays an important role in modifying the atmospheric circulation and climate over the surrounding area; it is an obstacle causing the westerly jet stream to be diverted to northern China. This strongly affects the development of monsoons in southern China and the aridity of western China. The climate of western China is strongly continental and may be among the most sensitive areas of the globe to possible feedbacks on the physical climate system from changes in terrestrial ecosystems.

Extensive anthropogenic changes in vegetation cover have occurred and are still occurring due to overgrazing, logging, irrigation, and conversion of marginal lands to agriculture. In addition to widespread land degradation and soil erosion, clear evidence from lake levels and climate records show that a drying and warming trend in the arid western parts of China has occurred in the past few decades. In contrast to the prevalent attitude in the West that the climate may be considered to be in a steady state, and, therefore, constant until proven otherwise, the prevailing attitude in China is that the climate is always changing, as has been amply recorded in China's long history.

## Research Highlights

The panel identified outstanding research projects on natural vegetation at the CAS Institute of Botany and one on agricultural systems at the CAS Shanghai Institute of Plant Physiology.

### *CAS Institute of Botany*

Under the leadership of Zhang Xinshi, director of the CAS Institute of Botany, very sophisticated information systems used to study climate–vegetation interactions in China have been developed at the Laboratory for Quantitative Vegetation Analysis. Databases on climate, topography, soils, land use, and vegetation cover have been integrated into a geographical information system (GIS) developed at the institute. Advanced Very High Resolution Radiometer (AVHRR) vegetation index data received and processed by the National Satellite Meteorological Center are being used by institute researchers. The computer facilities are principally advanced microcomputers with modern software. Vegetation zonation has been analyzed by using several classification schemes that are widely used for world-wide comparison and study of climate–vegetation interactions, including Thornwaite's and Holdridge's classifications, Budyko's radiative dryness index, and potential annual net primary productivity (NPP). In addition, multivariate methods have been used to rank and classify climate zones according to their climatological and geographical parameters.

This work provides a very strong foundation for studies of the effects of climate and $CO_2$ change on terrestrial vegetation. Work is in progress to extend the use of remotely sensed vegetation index imagery in studies of vegetation dynamics and to use radiative transfer approaches to model NPP. Models of ecosystem physiology and coupling of physiological processes to the physical climate system are needed to link these studies to general circulation model (GCM) simulations of alternative climate scenarios and to examine continental-scale vegetation feedbacks on the climate system. The technological level of the facilities at the CAS Institute of Botany, especially the computer equipment, is not high by Western standards. Nevertheless, the creative use of resources and intellectual sophistication of the approach make this a world-class effort.

Progress at the CAS Institute of Botany has been benefitted by strong international collaboration, including with Mark Harwell at the University of Miami and Alan Robock at the University of Maryland. Moreover, it appears that collaboration on climate–vegetation interactions has been easier to initiate with foreign scientists than with Chinese colleagues at other local institutions.

### CAS Shanghai Institute of Plant Physiology

Extensive and sophisticated work on agricultural ecosystems was identified at the CAS Shanghai Institute of Plant Physiology, which provides valuable foundations for further studies of the effects of climate and $CO_2$ change on agricultural ecosystems. Wang Tianduo is the lead scientist for a large-scale integrated study of the efficiency of water use for agricultural production on the North China Plain. This unique study takes an integrative view of the hydrology, irrigation management, cropping strategies, and economic factors that govern agricultural output for the lower Huanghe River system (Wang 1990). Of particular note is the regional focus of this research and the strong mechanical approach to modeling water use in the physiological processes of crops based on models developed by scientists at the institute. However, this dynamic and relevant project has received only minor funding under the Eighth 5-Year Plan. Wang Tianduo will be involved in a NSFC-funded project, "Water-Saving Agriculture on the North China Plain," which is scheduled to begin in 1992, although it does not appear that this project will be a major vehicle for continuing Wang's water use efficiency studies.

It would be possible to use the design of this study to examine the impacts of $CO_2$ fertilization on agriculture in a water-limited system or to examine the impacts of changes in precipitation or temperature of the region. This study also presents the opportunity to couple present physiological models to climate models in order to examine feedbacks from changes in the agricultural system on the regional climate. However, no such activities are planned at this time. This appears to be a unique study that presents the potential for examining the responses to $CO_2$ and to climate change of one of the most important and populous agroecosystems in the Earth system.

## Summary of GCTE Research

The diverse topography and climate of China provide unique opportunities to examine the role of climate and the effect of climate change on terrestrial ecosystems. The panel identified outstanding and relevant research projects on natural vegetation at the CAS Institute of Botany and one on agricultural systems at the CAS Shanghai Institute of Plant Physiology. However, Chinese research appears to place less emphasis on detecting evidence for change; instead, more emphasis is placed on impacts of and responses to change.

## BIOSPHERIC ASPECTS OF THE HYDROLOGICAL CYCLE AND GLOBAL ENERGY AND WATER CYCLE EXPERIMENT

Chinese researchers are active in modeling and measuring interactions between the land surface and the atmosphere. Significant efforts are ongoing in the development of models to represent the role of vegetation in controlling surface energy balance and evapotranspiration. The landscape of western China contains considerable contrast between well-watered irrigation districts along its rivers and surrounding arid areas. These produce effects on mesoscale atmospheric circulation and this phenomenon is the subject of both theoretical and empirical study, such as the Sino–Japanese experiment discussed below. Finally, because of China's historical dependence upon irrigation, hydrology is a mammoth enterprise, and considerable data on surface and groundwater hydrology exists at all scales, though these data are not—as yet—well integrated into the global change endeavor. Overall, China is poised to conduct significant work under the joint IGBP–WCRP project, Biospheric Aspects of the Hydrological Cycle and Global Energy and Water Cycle Experiment (BAHC/GEWEX). Below are details of active work in areas central to the BAHC/GEWEX research agenda.

### Research Highlights

#### *Sino–Japanese Atmosphere–Land Surface Processes Experiment*

The Sino–Japanese Atmosphere–Land Surface Processes Experiment (HEIFE [the Chinese abbreviation has been retained in English documentation]), a collaboration between CAS and Kyoto University, is a large study of land surface climatology and land surface hydrology in a region where intense irrigation has created an oasis with a strong atmospheric subsidence over the irrigated zone. The CAS Lanzhou Institute of Plateau Atmospheric Physics is the lead institution for this study, and other CAS institutions are involved in specific areas of investigation. The study will continue for about 5 years, with intensive field measurements in 1991 and 1992. These studies

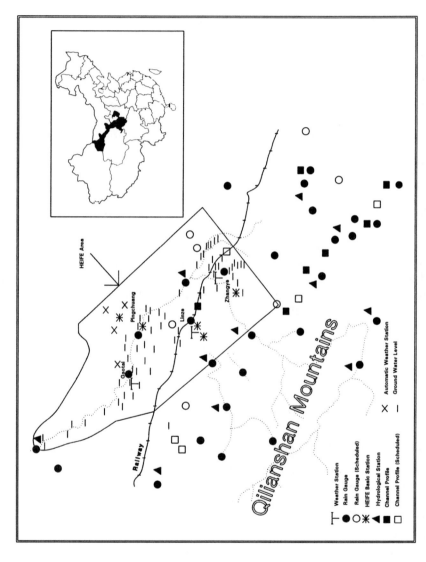

MAP 4-1 Sino–Japanese Atmosphere–Land Surface Processes Experiment (HEIFE [the Chinese abbreviation has been retained in English documentation]).

are to be applied to improving parameterization of fluxes in GCMs and development of water-saving irrigation procedures.

The site of the HEIFE experiment is an intensively irrigated region near Linze in Gansu Province (Map 4-1). The major water source is the Heihe River, which is fed by glaciers and summer (monsoon) snowfall in the Qilianshan mountain range. The irrigated oasis is surrounded by stone-surfaced (gobi) and shifting dune deserts. Rainfall in the HEIFE region is extremely low (119.5 mm w/l 3.5 mm snow annual precipitation). The elevation is 1,400 m, with a mean annual temperature of 7.6°C and a mean maximum temperature of 39.1°C.

Despite this harsh climate, wheat and corn production is high and some cotton and rice are grown. Wheat, corn, and soybeans are generally intercropped, with a wheat nurse crop providing sufficient humidity in its canopy to allow successful germination and early growth of corn or soybeans, despite prohibitively low humidity for those crops in the free atmosphere.

Irrigation management in this area is critically important, as salt water lies below the irrigation-derived water table and care must be taken to avoid salt contamination of shallow groundwater, thus the applied emphasis on management of evapotranspiration. Intensified irrigation in this part of the upper Heihe River has had an adverse effect on downstream water users in Inner Mongolia Autonomous Region, but this consequence does not appear to be a major consideration in water management around the experimental site.

*Chinese Research.* Work is being conducted in six major research areas:
- Analysis of turbulent structure and fluxes in the planetary and internal boundary layer
- Surface radiation fluxes
- Evaporation and surface water balance measurements
- Numerical modeling of boundary layer meteorology
- Development of techniques for water-conserving irrigation practices
- Development of a database of the region's land surface information for subsequent analysis and interpretation

*Japanese Research.* Work is being conducted in five major research areas:
- Large-scale circulations
- Water cycle in the Heihe River Basin
- Boundary layer dynamics
- Fluxes and transport of yellow sand and dust
- Water utilization and plant productivity

The field experiment involves continuous monitoring of basic climate parameters, surface radiation, groundwater level, stream flow, evapotranspiration (by using lysimetry), and crop biomass. In addition, four Intensive Observing Periods and Biological Observing Periods are scheduled. During both periods, additional soundings and eddy flux, atmospheric profiling, dust content, and $CO_2$ flux measurements will be conducted. During the biological observation periods, crop physiological parameters (crop height, leaf area index, transpiration, photosynthesis, leaf water potential, and root system mass) will be made. While no detailed information on sampling strategy for the biological parameters was made available, based on panel members' observations, it is likely that they will be made in a small number of agricultural test plots and not sampled in a spatially extensive fashion over the site. The experimental design for the study apparently was developed largely by meteorologists, and panel members saw little evidence of intellectual involvement by plant biologists or agronomists in this aspect of the projects' development.

The experimental design of the array of meteorological observation sites was decided through an innovative application of a numerical simulation model of the mesoscale circulation in the planetary boundary layer over the diverse surfaces included in the experiment. The area was sectioned into grids according to land use surface properties. Simulations were then conducted to examine whether potential meteorological sites provided accurate representations of their respective regions.

Use of remote sensing data is limited to the definition of land use over the area and is not available to define temporal dynamics of vegetation cover. Also, this project does not have an aircraft-based measurement component. These are clearly related to the project's limited funding and very high leasing costs for obtaining these services from other institutes or branches of the government.

It was difficult to ascertain the level of collaboration with the Japanese in this project, in part because the panel members' visit did not coincide with visits by any of these collaborators. Some of the equipment was clearly provided by the Japanese. However, it was also apparent that the Chinese input of expertise and equipment was significant. The major limitations appear to be related to the level of funding. Yet, the resources for some aspects of the project show abundant—if not redundant—use of resources, suggesting some difficulty in establishing priorities.

### Summary of BAHC/GEWEX Research

The HEIFE experiment is important in China as much for its role as a pathfinding study in interdisciplinary and integrated science as for the new findings concerning the oasis or cold island effect and other aspects of land–atmosphere coupling in arid regions. The study involves meteorological, ecological, hydrological, and agronomic measurements and atmospheric modeling. The project involves scientists from all of these disciplines, and, while integration is not perfect, it can serve as a model study.

The first prerequisite for the success of such a study—the acquisition of a high quality data set—clearly will be met. It is still unclear to what extent interdisciplinary analysis of the extensive data set will be completed, but workshops and data meetings are scheduled. The experiment has high visibility in China and its success can serve as a valuable model for future collaborative interdisciplinary projects.[5] While the experiment itself has been bilateral by agreement, scientists from other Chinese institutions and foreign scientists are welcome to participate in analyzing the data. In the future, multilateral rather than bilateral experiments under BAHC/GEWEX auspices will be a welcome and important next step.

## PROGRAMS ON MARINE ENVIRONMENTS

### Introduction

China has a coastline of over 18,000 km and claims more than 5,000 coastal islands. This coastline borders the inland Bohai Sea, Yellow Sea, East China Sea, and South China Sea, which collectively extend across temperate, subtropical, and tropical climate zones (Zhang 1984). Historical, contemporary, and future changes in land cover and basin hydrology will continue to alter drastically the delivery of water and sediments as well as eolian materials to the coastal zone, particularly in the deltaic regions of the Huanghe, Yangtze, and Pearl Rivers. These changes pose serious threats to the livelihoods of millions of people. The rapid development of mariculture in China and elsewhere in Asia has compounded these problems. While mariculture is a very efficient means of producing animal protein, it sometimes is detrimental to natural coastal environments. For example, it can threaten the sustainability of fisheries production offshore and cause pollution near shore.

Thus, China has considerable reason to be interested in the ocean, particularly in the coastal zone. This interest is reflected in the extensive activities China has undertaken since 1950. According to Zhang Haifeng (1984), a total of about 100 marine scientific and technological institutions are active in China and enjoy extensive communication with international marine organizations. Internal communication is promoted by a common national publishing house, the China Ocean Press.

Most marine research relevant to global change is carried out in CAS, which has the strongest basic marine science programs, as well as SOA, universities, and MOA. A list of leading marine organizations and their locales is provided below (Zhang 1984). Responsible administrative systems are identified in parentheses.

    China Ocean Press, (SOA), Beijing
    Dalian Fisheries College, Ministry of Agriculture (MOA), Dalian
    Department of Geo and Ocean Sciences, Nanjing University (SEDC), Nanjing
    Department of Marine Geology, Tongji University (SEDC), Shanghai
    Department of Oceanography, Xiamen University (SEDC), Xiamen
    East China Sea Fisheries Institute, Chinese Academy of Fisheries Science (MOA), Shanghai
    First Institute of Oceanography (SOA), Qingdao
    Institute of Acoustics (CAS), Beijing
    Institute of Coastal and Ocean Engineering, East China College of Water Conservancy (SEDC), Nanjing
    Institute of Marine Environmental Protection (SOA), Qingdao
    Institute of Marine Geology, Tongji University (SEDC), Shanghai
    Institute of Marine Scientific and Technological Information, National Oceanographic Data Center (SOA), Tianjin
    Institute of Ocean Engineering, Dalian College of Technology (SEDC), Dalian
    Institute of Ocean Technology (SOA), Tianjin
    Institute of Subtropical Oceanography, Xiamen University (SEDC), Xiamen
    Marine Fisheries Research Institute of Zhejiang Province (Zhejiang Province), Shenjiamen
    Marine Geomorphology and Sedimentology Research Division, Nanjing University (SEDC), Nanjing
    Nanjing Water Research Academy (Ministry of Water Resources and Ministry of Communications), Nanjing
    National Marine Environmental Forecasting Center (SOA), Beijing
    Ningbo Oceanography School (SOA), Ningbo
    Qingdao Institute of Oceanology (CAS), Qingdao
    Qingdao Ocean University (SEDC), Qingdao
    Research Division of Marine Economic Geography, Geography Department, Liaoning Normal University (SEDC), Dalian
    Second Institute of Oceanography (SOA), Hangzhou
    Shandong Institute of Oceanographic Instrumentation, Qingdao

Shanghai Fisheries College (MOA), Shanghai
South China Sea Fisheries Research Institute, Chinese Academy of Fisheries Science (MOA), Guangzhou
South China Sea Geological Survey, Ministry of Geology and Mineral Resources), Guangzhou
South China Sea Institute of Oceanology (CAS), Guangzhou
State Oceanographic Administration, Beijing
Third Institute of Oceanography (SOA), Fujian
Xiamen Fisheries College (MOA), Xiamen
Yellow Sea Fisheries Research Institute, Chinese Academy of Fisheries Science (MOA), Qingdao
Zhanjiang Fisheries College (MOA), Zhanjiang
Zhejiang Fisheries College (MOA), Zhejiang
Zhejiang Provincial Institute of Estuarine and Coastal Engineering Research (Zhejiang Province), Zhejiang

## Land–Ocean Interactions in the Coastal Zone

In late 1989, IGBP organizers recognized that many agents of global change would have specific impacts on the coastal zone. These agents include climate change, sea-level rise, and land use as it influences delivery of materials from the land to the sea. In addition to these new or forecasted impacts, the coastal zone already suffers from several aspects of intensive human use, such as overexploitation of fishing stocks, habitat destruction, and increasing concentrations of toxic substances. Such human effects, while not global in extent, are highly pervasive throughout the temperate and tropical coastal zones of the world. Recognition of this critical interface between land and sea came too late to permit development of a plan that could be accepted as an initial core project in 1990. Instead, Land–Ocean Interactions in the Coastal Zone (LOICZ) was presented as an initial concept for a core project in IGBP Report No. 12 (1990). This project is still in the planning stage, and its evaluation and acceptance as a core project is not expected before late 1992.

An IGBP working group held a NATO-sponsored workshop in October 1991 attended by approximately 40 coastal zone specialists who further refined research plans for LOICZ. The LOICZ–NATO workshop allowed scientists to fill gaps left by the initial definition of LOICZ and to identify three focal areas for LOICZ research: (1) land–ocean exchange, including activities on drainage basin dynamics, estuarine dynamics, atmospheric exchange and cross-shelf exchange between estuaries and coastal oceans with the open ocean; (2) biogeomorphology, including activities on sea-level change, coastal circulation, and physical feedback of biotic communities on coastal geomorphology; and (3) ecosystems and biogeochemical cycles in the coastal ocean, including activities in the biological structure of ecosystems, biological energetics and material processing (especially carbon, nitrogen, and phosphorus cycles), gas exchange, carbonate dynamics, and sediment dynamics.

It is important for this discussion of Chinese activities in marine science to note that the initial working definition of the coastal zone held by LOICZ was "the area extending from the landward margin affected by salt water to the outer edge of the continental shelf." At the same time, the Joint Global Ocean Flux Study (JGOFS) is dedicated to understanding the fluxes of materials in the open oceans beyond the shelf breaks (IGBP 1990). Thus, the shelf break, a potentially critical zone of hydrodynamics and biogeochemistry, was omitted in this planning. More recently, JGOFS and LOICZ have resolved to examine this interface zone jointly.

Although JGOFS is dedicated to understanding the fluxes of materials in the open oceans beyond the shelf breaks, it is clear that China is conducting a number of important research projects under the rubric of JGOFS that actually fit under the aegis of the proposed LOICZ, since all of the East China Sea and South China Sea are on coastal shelves. In fact, some of these Chinese JGOFS projects fit exactly with the kinds of activities planned by the latest LOICZ–NATO workshop. LOICZ organizers should definitely take into account the programs currently listed by CNCIGBP as JGOFS.

### *LOICZ Research Highlights*

Among marine research projects CAS considers part of China's global change program (CAS 1991), many of the studies would be important in understanding and predicting change in delivery of water and sediments to the coastal zone from land and some on the coastal seas themselves. The CAS Lanzhou Institute of Desert Research has studied the effects of climate change and land use change on desert riparian systems. The CAS Institute of Geography's Department of Hydrology has studied changes in land use and water and sediment delivery to the Huanghe and Yangtze Rivers. The CAS Nanjing Institute of Geography and Limnology has studied riverine hydrology modeling, coastal zone geomorphology, and historical analysis of land use change. The CAS South China Sea Institute of Oceanology has studied coastal and deep sea oceanography, estuarine processes, and coral reef growth rates. The CAS Qingdao Institute of Oceanology has recently begun a project on carbon cycling.

Outside of CAS, notable work was identified at two institutions. The Department of Geo and Ocean Sciences at Nanjing University has researched coastal zone geomorphology, delta subsidence, historical changes in land use, and sediment delivery by major rivers. And, the Guangzhou Institute of Geography has researched the historical geomorphology and interactions of sea-level rise and subsidence of the Pearl River Delta.

The NSFC has listed a number of supported programs in "Strengthening Coordination for more Effectively Funding for IGBP" (Zhang 1991), which is found in Appendix B. This document conveniently is organized by IGBP core projects, and specifically listed under LOICZ are seven very important projects, including sea-level change, flooding patterns, red tide, and changes in the Huanghe River drainage.

In combination, various institutions and universities appear to have coastal zone geomorphology well covered, and the potential for substantive contributions by China to LOICZ is excellent. These prospects would be even brighter if coordination among these institutions were strengthened and predictive work given higher priority.

## Joint Global Ocean Flux Study

The Joint Global Ocean Flux Study (JGOFS) has the following objectives: (a) the determination and understanding of "processes controlling the time varying fluxes of carbon and associated biogenic elements in the ocean, and to evaluate the related exchanges with the atmosphere, sea floor, and continental boundaries;" and (b) development of "a capability to predict on a global scale the response of oceanic biogeochemical processes to anthropogenic perturbations, in particular, those related to climate change" (IGBP 1990).

### *Chinese JGOFS Committee*

Chinese involvement in JGOFS began in February 1987, following the International Scientific Planning and Coordination Meeting for Global Ocean Flux Studies held at the ICSU Headquarters in Paris. The Chinese JGOFS Committee was established in February 1989, and in August 1989, the committee published the "Chinese Tentative Plan for the Joint Global Ocean Flux Study." This report reviews Chinese interest in JGOFS, provides scientific background for their activities, and outlines a tentative plan. An outline of future activity in JGOFS and a discussion of its overlap with the WCRP World Ocean Circulation Experiment (WOCE) is included in the Chinese JGOFS Committee's report. NSFC shares the costs of JGOFS projects ($1 million/5 years) equally with CAS (NSFC 1991a).

*National Program on Margin Flux in the East China Sea.* The Chinese JGOFS Committee's activities in the Eighth 5-Year Plan focus on the establishment of a national program on Margin Flux in the East China Sea (MFECS). Hu Dunxin, of the CAS Qingdao Institute of Oceanology, heads a CNCIGBP working group on ocean flux studies. The driving force of MFECS is the knowledge that coastal and marginal seas account for a significant portion of the primary productivity in the oceans. The East China Sea has one of the widest continental shelves in the world, receiving and accommodating a huge amount of material from the Yangtze River ($5 \times 10^9$ tons/year) and the Huanghe River ($10.7 \times 10^9$ tons/year), which are the two largest rivers emptying into the Pacific Ocean. MFECS has the following objectives:

- Understand the contribution of the seas in the northwest Pacific margin, for example the East China and Yellow Seas, to the flux or budget of carbon and relevant biogenic elements of the Pacific Ocean.
- Understand the oceanic processes (physical, biological, chemical, and sedimentological) controlling the margin flux in the study area.
- Provide validated flux numbers for carbon and related elements between the Pacific Ocean and the East China Sea.
- Assess the primary productivity in the East China and Yellow Seas in order to provide scientific background for assessing biological productivity and resources.

MFECS will address the following scientific questions:

- Is primary production in the East China Sea supported by nutrient input from the Pacific Ocean or from the Yangtze River? What processes control this input?
- Are the organic materials produced in the East China Sea exported seaward to the Pacific Ocean by the Kuroshio Ocean Current or landward to the near-shore zone by upwelling?
- Is the Okinawa Trough a sink for carbon and other materials? What processes control the lateral exchange of carbon and other materials between the Pacific Ocean and the East China Sea?
- What are the processes controlling the vertical flux of carbon and the related biogenic elements in the East China Sea?

These questions will be answered by carrying out the following research:

- Study the vertical flux processes in some typical areas of the Yellow and East China Seas by studying the air–sea exchange mechanism: $CO_2$ fixation by phytoplankton and transport of carbon in the euphotic zone and the biogeochemical and sedimentological processes in benthic layer.
- Study the horizontal flux processes in the East China Sea margin by examining horizontal fluxes of water and other material through physical processes near the Kuroshio Ocean Current.
- Study the transport of carbon and other materials on the East China Sea shelf by examining the input of carbon and other material from the Huanghe and Yangtze Rivers.
- Study the dynamics of the transport of carbon and other material on the shelf of the Yellow Sea and the East China Sea.
- Study the role of (terrestrial) calcium carbonate in the carbon cycle in the southern Yellow Sea and northern East China Sea.
- Study the dynamic processes of sediment in the Okinawa Trough.
- Study remote sensed data in combination with *in situ* measurements of phytoplankton.
- Model vertical flux processes in typical areas of the East China Sea and the Yellow Sea, including modeling of margin flux and the role of physical and biogeochemical processes.

An initial MFECS activity, an international workshop on the margin flux in the western boundary area of the Pacific Ocean was scheduled to be held in mid-June 1992 in Qingdao. Following the conference, a Chinese field experiment is planned for late 1992. This experiment would be followed by an international field experiment in the second half of 1993.

## Tropical Ocean and Global Atmosphere

The Chinese are participating in the WCRP Tropical Ocean and Global Atmosphere (TOGA) Core Project, which is devoted to understanding the exchange of energy between the atmosphere and tropical waters. The three principal organizations involved in this research are SOA, SMA, and CAS. The Chinese have studied the so-called "warm pool" of the western Pacific Ocean. Under the "Agreement on U.S.-PRC Cooperation in TOGA Coupled Ocean–Atmospheric Response Experiment (TOGA/COARE)" (a more specific short-term project), the Chinese will carry out a coordinated set of air–sea measurements to be made in the South Pacific between 1 November 1992 and 28 February 1993.

### Observation Programs

Activities include a joint U.S.–China program between NOAA and SMA involving air and sea interactions in the tropical West Pacific. From December 1985 to 1990, one ship completed eight cruises, and data from these expeditions were published in January 1992 in a book of common papers. A second program, the Australian Monsoon Experiment, is a trilateral U.S.–China–Australia program. The third CAS observation program, "Air–Sea Interaction in the Tropical West Pacific and Interannual Variability of Climate," operated from 1985 to 1989.

### TOGA-Monsoon Climate Research

Monsoon climate research involves CAS, SMA, and selected universities and focuses on eastern Asian monsoons. One cooperative program between SMA and the National Science Foundation (NSF), "PRC–U.S. Cooperative Studies on Eastern Asian Monsoons," began in 1983 and will terminate in 1992. A monograph, *Eastern Asian Monsoons*, is in press.

### Monitoring of El Niño Events

This aspect of Chinese involvement in TOGA began in 1986 and is coordinated by the SMA. As part of this program, the *Climate Monitoring Bulletin* has been published monthly since 1989. The SOA, under the US–PRC Protocol on Cooperation in Marine and Fishery Science and Technology, and CAS as a group under the Chinese National Climate Committee (CNCC), are both studying the air–sea interaction directly in accordance with the TOGA project. Even though SMA and universities are involved in TOGA, the information presented to the panel predominantly described CAS activities.

From 1985 to 1990, nine expedition cruises (each lasting for 2 months) were conducted during winter months in the tropical ($25°$ N and $10°$ S) western Pacific (west of $150°$ E) by two vessels, Science No. 1 and Experiment No. 3. The cruises focused on the fluxes of momentum, heat, moisture, and other materials at the air–sea interface by using approximately 20 kinds of measuring equipment. Four billion data have been indexed and stored in a data bank that can be used for air–sea interaction studies, ground-truthing, and other research purposes. Following the PRC–USA Bilateral Coordinating Group Meeting on TOGA/COARE cooperation held in Guangzhou early in September 1991, CAS research vessels Science No. 1 and Experiment

No. 3 and SOA research vessel *Xiangyanghong* No. 5 were chosen to participate in TOGA.

The Chinese have tried to forecast *El Niño* and *La Niña* events by using historical and observational data collected from 1985 to 1990. The 1987-1988 *El Niño* and the 1988-1989 *La Niña* forecasts were successful, but the 1988-1989 *El Niño* development was not. In the process, the Chinese discovered different types of *El Niño*, *El Niño* development processes, and climate effects over China. As it turned out, the relationship between climate over China and *El Niño* was not a simple statistical problem.

For basic research on air–sea interaction, the Chinese are undertaking various activities: (1) different methods of measuring fluxes; (2) an air–sea column heat budget study by using data from two-vessel synchronous observation; (3) publication of the *Atlas of Climate Physics of Tropical Pacific Ocean*; (4) a newly designed air–sea coupled numerical model employing the daily synchronous coupling technique; and (5) a laboratory study of the air–sea interface, which has shown that the "low of wall" obtained from rigid surface is not valid for large areas over the ocean.

In the Eighth 5-Year Plan, NSFC and CAS are each contributing 5.5 million *yuan* ($1 million) and SOA 55 million *yuan* ($10 million) to fund TOGA. NSFC also supports WOCE but amounts and proportions were not given (NSFC 1991b).

## Summary of Research in Marine Environments

China is located on a large piece of the world's coastal land area and a substantial scientific infrastructure addresses coastal zone issues in a first-rate manner. The Chinese JGOFS plans (which, under current definitions, can be considered LOICZ activities) and TOGA involvements are ambitious and valuable components of IGBP and WCRP. The participating scientists are well qualified and eager for international cooperation. In addition to marine research, China offers a combination of land-based research endeavors that are very important to a LOICZ focus on land-ocean interactions (Shi et al. 1990). It is quite likely that Chinese efforts would be restricted to Chinese coastal zones—whether in cooperation with TOGA or LOICZ—unless activities were funded from international sources. As LOICZ planning continues, involvement of Chinese scientific leadership might bring particular strength to activities in coastal Asia—an area identified by LOICZ planners as being especially interesting and important from a global perspective.

Plans described by the CNCIGBP for marine research in the national global change plan are ambiguous. The January 1990 status report of the CNCIGBP (CNCIGBP 1990a), has a section under supporting activities titled "Ocean Studies in the China Sea in Association with Global Change." In this section eight studies are listed that would be associated with LOICZ objectives, several of which may or may not involve JGOFS objectives for deep ocean studies, and several of which fit into the TOGA program. Notably, much of the Chinese involvement in TOGA is on the broad marginal shelves of the China coast rather than over deep water. In addition, other land-based projects are listed throughout the report that would be important to the land-to-sea component of LOICZ. The September 1990 status report of the CNCIGBP (1990b), indicates a focus on changing sea levels under GCTE activities and some activities listed under JGOFS that are largely coastal rather than deep sea activities.

## GLOBAL ANALYSIS, INTERPRETATION, AND MODELING

The Global Analysis, Interpretation, and Modeling (GAIM) aspect of IGBP was initially envisioned as a core project, but in 1992, the Scientific Committee of the IGBP decided that its cross-cutting nature would be better served if it was administered as a coordinating committee. GAIM encompasses the broad area of development and analysis of coupled atmospheric and oceanic GCMs and of models of land ecosystems, along with analysis and interpretation of global data sets. Priorities for GAIM focus on global biogeochemical modeling coupled to climate system modeling. Primary responsibility for the international continuation of global climate model development lies with WCRP, however. Thus, key activities for GAIM are modeling terrestrial and marine primary productivity, carbon storage, and trace gas production. Transport of chemical substances in the atmosphere and oceans is also central to GAIM's mission.

Chinese IGBP activities contain relatively few efforts aimed directly at the central GAIM questions. However, many activities contribute materially to GAIM, which are described elsewhere in this report under IGBP and WCRP core program descriptions. The reader is referred to work on ecosystem dynamics described under GCTE, the CAS Institute of Botany section in Appendix A, the report on land process modeling under BAHC/GEWEX, the report on marine modeling under JGOFS and LOICZ, and the report on climate modeling in the CAS Institute of Atmospheric Physics. As GAIM evolves internationally, it is likely that Chinese participation will increase, and involvement of Chinese scientists should be encouraged so that activities already under way can be further developed.

## SYSTEM FOR ANALYSIS, RESEARCH, AND TRAINING

The System for Analysis, Research, and Training (START) initiative provides the infrastructure for regional research and cooperation in support of the IGBP, WCRP, and the ISSC/IGBP Human Dimensions of Global Environmental Change (HD/GEC) programs. The START initiative was launched at a meeting sponsored by the ICSU Scientific Committee for the IGBP (SC–IGBP) in December 1990 in Bellagio, Italy, where participants found that "the most effective configuration of regional collaboration would be a global system of regional networks dedicated to analysis, research, and training." Each regional research network (RRN) would consist of a number of regional research sites (RRS) and a regional research center (RRC) (IGBP 1991). In December 1991, the START Standing Committee,[6] issued guidelines for the establishment of RRCs, including objectives,[7] functions, and criteria for regional START proposals.

The START Standing Committee has identified 13 approximate geographic regions that "span a scientifically coherent area," and China has land mass in three: Central Arid Asia, Tropical Asian Monsoon, and Temperate East Asia.[8] The START Standing Committee has received funding from the Global Environment Facility (GEF)[9] to establish RRNs for the Tropical Asian Monsoon Region in Association for Southeast Asian Nations (ASEAN) countries,[10] for the Northern African Region in Africa north of the Equator, and for the Inter-American Institute for Global Change Research[11] (IGBP 1992a, 1992b).

### Chinese START Proposal

The CNCIGBP is developing a proposal to the START Standing Committee to establish a RRN for East Asia and Western Pacific (EAWEP) and to house its RRC at CAS.[12] While this version of the proposal does not yet meet newly established START guidelines for establishing RRNs, it does show that CAS does have substantial resources to contribute to START objectives.

*Regional Geography*

The region for the proposed network would include Japan, Korea, the Philippines, Republic of Mongolia, the far eastern part of the former USSR, and some Pacific Islands. The proposal does not yet outline how multilateral work would be organized and administered.

### Scientific Themes

The Chinese propose four central scientific themes for the EAWEP network.

*Impact of global change on sustainable development.* This theme focuses firstly on regional climate and environmental responses to global warming, including impacts of warming on the monsoon system, sea level, water resources, and ecosystem structure and function. A second area of study is the impact of global change on economic development, in particular the impact on agriculture in coastal zones (including islands). The third area of study is the impact on human society.

*Regional problems of global significance.* Five research areas are put forward: (1) the role of the changing monsoon system in the global hydrological cycle and global climate; (2) methane from rice paddy production; (3) changing land use patterns, for example, the deforestation and renewability of tropical monsoon forests; (4) land–ocean interactions in the coastal zone; and (5) past global change studies from records of the Qinghai–Tibet and Loess Plateaux, deep sea cores, and proxy data from Chinese literature.

*Development of human resources.* The Chinese want to establish an international school for global change to address all types of educational and training needs (Chapter 3). They have approached the Pacific Science Association, which currently is headed by CAS President Zhou Guangzhao, for assistance in funding a fellowship program for global change education.

*Policy and strategies for global environmental issues.* This theme was not very well articulated in the March version of the proposal. No specific approaches or policy areas were identified, although mention was made of promoting sustainable development and North–South cooperation on global environmental issues.

### Network Organization

In general, the Chinese proposal follows the organizational and operational outlines recommended for START at the Bellagio meeting. One modification is the proposed subcenter for Pacific Islands, located on an unnamed Pacific island, that would assist the RRC by dealing specifically with issues particular to those islands.

The EAWEP RRC would be located in the National Center for Global Change Research, which CAS is proposing to establish, and which would be administered by a secretariat. CAS would host the RRC because of its leadership role in basic sciences and because of its strong scientific infrastructure. CAS also hosts the China Center for the ICSU Committee on Data for Science and Technology (CODATA) and is the secretariat for the ICSU World Data Center (WDC)-D, and five of the nine WDC-D subcenters[13] are in CAS. Furthermore, CAS is in the process of upgrading its ecological stations into a network for long-term studies.

### Summary of the CNCIGBP Proposal

The draft CNCIGBP proposal to the START Standing Committee demonstrates the large and multidisciplinary research enterprise that CAS offers to the study of environmental and climate change. The proposed scientific themes closely reflect the current Chinese research agenda. Overall, this proposal will be further strengthened when details are added concerning multilateral links in the proposed EAWEP region.

Of the developing countries, China offers one of the most advanced scientific infrastructures. Furthermore, it is an important site for research in many of the global change topics. A strong role for China in START would be beneficial to China, the region, and the three major international global change programs.

## DATA AND INFORMATION SYSTEMS FOR THE IGBP

Data and Information Systems (DIS) for the IGBP was established to provide global data needed by core projects and, eventually, to provide data management and information services (IGBP 1990). Since its inception, DIS has been open to China's active involvement, for example, in the development of a 1 km global AVHRR data set and in having a site for one of the DIS land cover change pilot studies.

Under the leadership of Fu Congbin, who is from the CAS Institute of Atmospheric Physics and a representative from China to the SC-IGBP, the Chinese global change program has two relevant activities using Chinese AVHRR data. First, Fu has used AVHRR data to produce a normalized difference vegetation index (NDVI) for the country. Second, and most recently, in collaboration with the State Meteorological Satellite Center on a national project of the Eighth 5-Year Plan, the CAS Institute of Atmospheric Physics will produce 1 km AVHRR data sets for the first time in China.

In IGBP Report No. 12 (1990), listed under DIS are various sites around the world for land cover change pilot studies, one of them being the Gansu Grassland Project. In 1990, the Gansu Grassland Ecological Research Institute was named as the lead institute for a national grasslands monitoring project to be funded by MOA. At the time, it seemed that this project, headed by Chen Quangong, would fit nicely into the DIS agenda. However, attempts to have Chen attend IGBP Land Cover Change Working Group meetings have not been successful.

At the IGBP Asian Planning Meeting in December 1991, one of the DIS recommendations would promote linkage between DIS and regional and national centers responsible for data (IGBP 1992a). China has national data centers at CAS, SMA, and NEPA that could be more strongly linked to DIS activities.

Data management and accessibility are very important issues in science. In China, data often are not well managed and can be considered commodities at the institutional and agency levels, and, therefore, are not easily available. Chinese participation in international programs would prompt better management and improve data accessibility. While DIS will primarily address global data needs in support of core projects, the Chinese can pursue their data and information needs and interests by increasing their participation in programs such as START and GAIM.

## HUMAN DIMENSIONS OF GLOBAL ENVIRONMENTAL CHANGE PROGRAM

According to NSFC, China plans to participate in the Human Dimensions of Global Environmental Change (HD/GEC) Program that is sponsored by the International Social Science Council. The goals of the HD/GEC program are consistent with China's research priorities; in fact, the study of the impact of human activities is evident to one degree or another in many current research agendas. However, this area of the global change research agenda often requires interdisciplinary research, which challenges the way most Chinese research is organized (Chapter 3).

In the mid-1980s, CAS carried out some research that has resulted in some published reports and articles: "Survival and Development," "On a Sustained and Harmonious Development of National Economy," "Current Status, Causes, and Strategies of China's Ecological Environment," and "Study of Chinese Population Development." The purpose of these works was to report on the interactions between humans, natural resources, and development.

The CAS Institute of Geography has compiled a map of land resources in China (1:1,000,000) and national, regional, provincial, and county maps of land use in China (1:1,000,000) that will be important to research on land use change and human dimensions of global change.

## Research Highlights

An example of the type of research that the Chinese are considering under the human dimensions rubric is a CAS project, "Analysis in China's National Conditions: Research on Harmonious Development of China's Population, Resources, Environment, and Economy," led by Zhou Lisan, CAS Nanjing Institute of Geography and Limnology. Also participating are the Commission for Integrated Survey of Natural Resources (CISNAR), RCEES, and the CAS Institute of Systems Science. This 2-year project is scheduled for completion in 1992. The study will examine population, resources, environment, economies, and differences in regional development in order to develop long-term strategies for overall economic development. Additional examples of projects can be found in Appendix B.

## Summary of HD/GEC Research

The highly applied nature of much of China's science fosters and often demands that relationships between man and nature be investigated. Unfortunately, like most countries, China's scientific and educational infrastructures and funding mechanisms are not organized in a way to promote substantive and prolonged interdisciplinary work. Still, given China's resource consumption patterns, natural resource base, population growth, and settlement patterns, it probably stands to gain more than most countries from actions to promote the HD/GEC agenda and, with it, the justification and basis for interdisciplinary work.

## CHINESE ECOLOGICAL RESEARCH NETWORK

The IGBP has provided further justification of and context for long-term ecological research, and CAS has capitalized on this link in promoting its Chinese Ecological Research Network (CERN) as a distinct component of China's national global change program. As it is recognized in the IGBP, the value of global change research will depend on a structure allowing data to be shared and analysis to be carried out at different scales. The plan for CERN seeks to establish such a structure for Chinese ecological data and analysis.

CAS first started to organize CERN in 1987 to reorganize and improve the way ecological research and data collection could be undertaken at its 52 ecological and monitoring stations located throughout the country. An internal review of ecological work showed a lack of standardization of methodologies, data management, and of requisite equipment (Zhao 1990).

While CERN's mandate supports global change research, its main mission supports governmental national resource development policies, which seek the maximum exploitation of natural resources. The CERN research program emphasizes local and regional human impacts on the environment, improved agricultural production, resource management and development, and long-term environmental monitoring.

## Organization of the Network

Over the next five years, CAS will implement CERN by upgrading 29 of its 52 ecological research and monitoring stations. In the Eighth 5-Year Plan, CAS has budgeted 45 million *yuan* ($8.2 million) for construction and equipment and 10 million *yuan* ($1.8 million) for CERN-related research. In addition, CAS is applying for a $16 million World Bank loan for equipment and training.

CERN is organized into three parts: (1) a network of demonstration sites, which would show land use and reclamation practices, low-input agricultural production techniques, and other "optimization" production techniques; (2) a research network, which would conduct studies on 10 core topics in the ecological sciences, atmospheric sciences, and agronomy; and (3) monitoring networks consisting firstly of a "basic observing system" that will measure meteorology, hydrology, and biology and at all 29 CERN stations, and secondly of a network that will measure trace gases, wet and dry deposition of heavy metals, and other variables at a limited number of stations.

### *Synthesis Center*

The Synthesis Center, which is in the planning phase, will be located at CISNAR. CISNAR already houses the Integrated Research Center for Natural Resources and Agricultural Development (established in 1990 and sponsored jointly with the National Commission for Agricultural Regional Planning) and World Data Center-D for Renewable Resources and Environment, as well as various other large-scale (regional and national) databases based on CISNAR survey projects over the years.

Synthesis Center research will focus on regional and national scales and interdisciplinary studies, with an emphasis on natural resource management and policy questions.

### *Subcenters*

Four institutes have been named to coordinate and validate data collected at CERN stations. These subcenters are organized by discipline: soil (CAS Nanjing Institute of Soil Science), hydrology (CAS Institute of Geography), atmosphere (CAS Institute of Atmospheric Physics), and botany and zoology (CAS Institute of Botany) (Appendix A). While the packaging of data along traditional disciplinary lines is not ideal for many global change issues, the subcenter institutes are centers of excellence for global change research. For example, the CAS Institute of Botany is a leader in ecological modeling and is well placed to develop variable-scale analyses. The subcenters will concentrate on disciplinary studies.

## *Field Stations*

Of the 29 CERN stations initially chosen to be included in the network, 11 have been designated "leading" stations. These stations have been selected based on their superior research and on the strength of their research staff, physical plant, and equipment relative to other stations. These stations will have expanded research and monitoring agendas and will be sites for remote sensing and regional monitoring studies. The remaining stations have been chosen based on criteria similar to those employed for "leading" stations. A selection process was employed in part because of administrative and financial considerations.

CAS has determined that each field station should have a minimum of five basic maps: topography, soil, vegetation, geomorphology, and remote sensing imagery. Beginning in 1992, CAS will initiate a plan to install one personal computer and GIS software at each field station in order to digitize these maps.

## CERN Information System

The key to CERN's success will be its ability to produce valid, well documented, and accessible data. CAS officials are well aware of the importance of improving research data management capabilities and of making data available nationally and internationally. To this end, CAS is investing heavily in data and information management over the next five years. And, eventual links to the CAS computer network will further improve the CERN Information System.

CERN planners have devised an index system of structure and function variables for ecosystems that will be the basis of six databases to be set up. Database I will include basic ecological and environmental data that will be collected by all 29 CERN stations and sent to the Synthesis Center for analysis. Database II will contain process-oriented data collected at certain specified sites, for example, from the specialized research assigned to "leading" stations. To date, three sets of standardized methodologies (forest, agriculture, and grasslands), along with one book of international standard methodologies, have been produced to meet field station needs.

Database III and its subsets relate primarily to field stations: (1) background information on field stations, including natural environment, resources, and energy, (2) social and economic data (probably up to county level), (3) management data for individual research projects (at least metadata such as methodologies, main results, and principal investigator), and (4) station administration.

Database IV will contain data collected from other projects that may be of use to CERN. Database V will contain data on large-scale data bases collected by the four subcenters. Database VI will contain social and economic data.

Databases at the Synthesis Center will be considered permanent, which includes long-term monitoring data and data from structure and function research and demonstration projects. Social, economic, and natural resource data will also be permanent. Data from process-level studies are not considered permanent and will be managed by researchers until they are finished (in approximately 5 years) analyzing their results. Regardless, scientists will not be forced to relinquish data. Once the process or ecosystem study is complete, data will be stored at the subcenters or Synthesis Center, depending on the CERN research needs and the type of research undertaken.

## CERN Research

### Core Projects

The first experiment to be conducted under CERN auspices, the "Structure and Function of Major Ecological Systems in China and Demonstration of Their Optimum Managerial Models," was begun in 1989. Currently called the "Network Study on Ecosystems in China: Study on the Structure and Function of Main Ecosystems in China and Approaches to Increasing Their Productivity (1991-1995)," it outlines standardized research organized by four major ecosystems: agroecology, forest, grassland, and aquatic field stations. The study outline can be found in Appendix D. CERN's ten core projects are listed below:

- Study of the impact of climate change and human activities on ecosystem degradation
- Monitoring and prediction of natural disasters

- Study of the impact of climate change and human impacts on fragile ecotones
- Study of NPP on a national scale and its relationship to climate change and human activities
- Study of geographic divergence of water cycling and water balance
- Study of the geographic divergence of the energy cycle
- Study of the geographic divergence of biogeochemical cycling
- Study of crops
- Study of the rules of ecosystem succession
- Development and use of optimized management models

Within these projects, certain basic research will be conducted on such topics as those listed below:

- The hydrological cycle, including precipitation, surface water, soil moisture, and ground water models
- The nutrient cycle, including the prediction of soil nutritional levels and research on the role of soils in the carbon, nitrogen, sulfur, and phosphorus cycles
- Trace gases fluxes and their generation, transportation, and transformation in various ecosystems
- The decomposition, accumulation, and transport of organic chemical pollutants and heavy metal elements
- The energy flow process

## *Cooperative Links Between CERN and the U.S. Long-Term Ecological Research Network*

Chinese and American scientists in the U.S. Long-Term Ecological Research (LTER) Network have been involved in successful ecological cooperative projects both in the United States and in China, including coarse woody debris studies (H.J. Andrews LTER, Oregon State University), modeling (Virginia Coast Reserve, University of Virginia; Central Plains Experimental Range, Colorado State University), and data management (Sevilleta LTER, University of New Mexico; H.J. Andrews LTER). CERN stations involved in these types of collaboration include Changbaishan, Xilingele, and Dinghushan (Appendix D). Support for this work has come from various sources such as the National Science Foundation and UNESCO's Man and the Biosphere (MAB) Program. These links to American LTER scientists and sites have been major avenues for the introduction—on a project-by-project basis—of new research techniques and tools such as modeling and GIS, which have positively influenced the development of CERN (Leach 1990).

In 1990, CAS asked the CSCPRC to help strengthen these cooperative links. Specifically, CERN planners were interested expanding their knowledge of how LTER is organized and LTER data are managed. In May 1991, a CERN delegation led by Sun Honglie, CAS vice president and CERN chairman, visited five LTER stations and the LTER Network office. Of particular interest to the delegation was intersite cooperation, data management, network communications, and site management.[14] In September 1991, a delegation of LTER scientists, led by James Gosz (Sevilleta LTER), went to China, where they visited CERN stations in forest, grassland, desert, and agricultural ecosystems and discussed possibilities for cooperative studies at these sites (Gosz and Leach 1992).[14] While they were there, LTER scientists were also part of a World Bank team that reported to the Bank on the state of CERN plans (Leach et al. 1992).

### *Participation in the MAB Program*

Chinese participation in the MAB program is financed by CAS and has a secretariat of six persons located in the CAS Bureau of Resources and Environmental Sciences, which also administers CERN. The chairman of the China MAB Committee is Sun Honglie, who is also a CAS vice president, director of the Commission for Integrated Survey of Natural Resources, chairman of CERN, and a member of CNCIGBP. China MAB has a budget of about 130,000 *yuan* per year.

The Chinese MAB program has no particular mandate to fulfill any part of the global change research plan. The most important function is the maintenance of biosphere reserves that will be important sites for the monitoring of ecosystem changes. China has eight biosphere reserves, and four of them are also CERN sites: Changbai Mountain, Jilin Province (temperate forest, World Heritage Site, and a CERN site); Dinghu Mountain, Guandong Province (CERN site); Fanjing Mountain, Guizhou Province; Wuyi Mountain, Fujian Province; Tianchi Lake, Xinjiang Uighur Autonomous Region (glacial lake, alpine forest, and meadow and a CERN site); Wolong, Sichuan Province; Xilingele, Inner Mongolia Autonomous Region (grassland; CERN site); and Shennongjia, Hubei Province.

Like cooperative research with LTER scientists, the MAB program has been an important avenue for the introduction of new ecological concepts, techniques, and training. Examples of projects include the Cooperative Ecological Research Project, which was carried out with Hans Brunig of Hamburg University and others on ecosystem processes at the Xiaoliang station of the CAS South China Institute of Botany, an ecosystem restoration project with Sandra Brown at the

University of Illinois at the CAS South China Institute of Botany, and a project on the comparison of broadleaved forests at Changbai Mountain for which Orie Loucks, Miami University (Ohio), is the American PI.

## Summary of CERN

CERN is an ambitious and important commitment by CAS to improve the way it conducts ecological research. The large information and data management requirements will demand closer attention to quality assurance, documentation, and standards. The scope of this undertaking has significant implications for the types of contributions China can make to ecological studies and international scientific programs.

## NOTES

1. The four tasks of the DOE–CAS Joint Research on the Greenhouse Effect are carried out by scientists at the State University of New York at Albany and at Stony Brook, Lawrence Livermore National Laboratory, National Center for Atmospheric Research, National Oceanic and Atmospheric Administration, Oregon Graduate Institute of Science and Technology, and Oak Ridge National Laboratory on the U.S. side and by the CAS Institute of Atmospheric Physics and the CAS Institute of Geography on the Chinese side. See also details in Appendix C.

2. The Xi'an Laboratory of Loess and Quaternary Geology (Appendix A) and the University of Rhode Island are conducting a bilateral comparative study (Appendix C) on atmospheric transport of soils that is based on CHAASE research.

3. The Chinese did not participate in the aircraft experiments and many difficulties were encountered in trying to use China as a base for aircraft operations. After much negotiation, the NASA aircraft was allowed a short stopover in Shanghai and only on the condition that all scientific instruments be shut off. This is a good example of the type of challenges that can arise in bringing China fully into some of the international experimental programs.

4. Liu Tungsheng, director of the Xi'an Laboratory for Loess and Quaternary Geology, is a member of the PAGES Scientific Steering Committee.

5. Eric Smith, Florida State University, and Kuo Nan Liou, University of Utah are developing a land-surface climatology collaborative project with CAS. The experiment site would be at the Inner Mongolia Grassland Experiment Station, about 70 km south of Xilinhot, in a temperate, semi-arid continental steppe zone of typical grassland vegetation. Currently, plans call for experiments to begin in 1996. If funded and carried out, this project would complement the HEIFE experiment by contrasting water and energy fluxes in an irrigated environment with those fluxes in a dryland environment.

6. Li Wenhua, Commission for Integrated Survey of Natural Resources, is a member of the committee.

7. RRNs and RRCs should organize around five objectives: (1) a focus on *data* and information management, accessibility of data, data exchange with international programs, and coordination with IGBP DIS, WCRP, and HD/GEC data programs; (2)

*research, analysis, and modeling* to facilitate interdisciplinary research, analysis, and modeling at the regional level; (3) *policy outreach* to encourage the transfer of findings into policy, which will be accomplished in part by involving policy makers in network activities; (4) *training* to develop indigenous scientific capabilities through training, collaborative research, and scientific and technical cooperation; and (5) *scientific cooperation and access* through exchange and collaboration among RRNs and through dissemination of database directories and information about projects and network activities (IGBP 1992a).

8. Actual delineations of regional boundaries will be determined by "regional needs and desires, through discussions with appropriate representatives from the nations involved" (IGBP 1991).

9. The Global Environment Facility is a multilateral fund set up by governments, the World Bank, the United Nations Environment Program, and the United Nations Development Program to finance grants and low-interest loans to developing countries for projects related to global environment, for example, greenhouse gas response strategies, biodiversity action plans, and technology transfers.

10. China is not a member of ASEAN, and, consequently, will not receive funds from this particular GEF proposal to participate in Tropical Asian Monsoon regional efforts. However, China is welcome to participate in this region through other avenues, and these are being actively explored.

11. Original signatories to the agreement establishing the institute are Argentina, Bolivia, Brazil, Costa Rica, Dominican Republic, Mexico, Panama, Peru, United States, Uruguay, and Venezuela.

12. Information for this section is based on the March version of the draft "Proposal to the IGBP START Standing Committee to Establish a Global Change Regional Research Network for East Asia and Western Pacific Region" (CNCIGBP 1992). Since then, much progress has been made in developing and strengthening the proposal and in defining ways China can contribute to START. Discussions are ongoing with the START secretariat, including a recent visit to China by Thomas Rosswall, acting director of the secretariat.

13. Databases are maintained at the following WDC-D subcenters: earthquake data at the Department of Science and Technology, State Seismology Bureau; oceanography data at the Institute of Marine Scientific and Technological Information, SOA; atmospheric data at the Information Office, National Meteorological Center, SMA; geology data at the Institute of Geology, Chinese Academy of Geological Sciences, Ministry of Geology and Mineral Resources; renewable resources and the environment data at CISNAR, CAS; astronomy data at the Beijing Astronomical Observatory, CAS; glaciology and geocryology data at the Lanzhou Institute of Glaciology and Geocryology, CAS; geophysical data at the Institute of Geophysics, CAS; and space science data at the Research Center for Space Science and Applications.

14. These exchanges were organized in close cooperation with the U.S. LTER Network. The CERN delegation's visit was jointly funded by the NSF U.S.-China Program and the NSF LTER Program.

# 5

# Selected Topics

## INTRODUCTION

The panel was asked to select a limited number of disciplines or research areas that complement specific components of the U.S. national global change research agenda and the International Geosphere–Biosphere Programs (IGBP) core projects and to report on these topics in greater detail. Specifically, the panel was asked to report on Chinese capabilities, policy commitments, and potential for collaboration with the U.S. scientific community and for contributions to international research efforts.

The panel chose two focal areas: atmospheric chemistry and physical and ecological interactions of the atmosphere and land surface. The topics in the atmospheric chemistry focal area are very narrowly defined, while the topics in the second focal area are necessarily more broadly defined. Within these areas, specific topics were chosen by the panel based on members' expertise and on panel opinion about the relevance to U.S. and international research.

It should be noted that no attempt was made to be comprehensive in examining all of the possible topics available for discussion in a given focal area. For example, even though biogeochemistry includes more than trace gas research, the panel chose to limit its detailed investigation to biotic controls on selected trace gases, climate–vegetation dynamics, and soil organic matter and soil nutrient turnover.

## ATMOSPHERIC CHEMISTRY

Coal accounts for approximately 75 percent of China's annual energy consumption. Emissions of particulate matter and sulfur dioxide ($SO_2$) from burning coal are major contributors to regional air pollution. These emissions not only lead to urban and regional pollution problems such as oxidants and acid precipitation, but potentially also have global impacts. Anthropogenic emissions of ozone ($O_3$) precursors, such as nitrogen oxides ($NO_x$) and hydrocarbons, can lead to a significant increase in tropospheric $O_3$ (IPCC 1990). For example, $O_3$ is known to contribute significantly to the infrared radiation in the upper troposphere and, therefore, plays an important role in climate change (IPCC 1990). Remarkably high levels of tropospheric $O_3$ over northeastern China and Japan in spring and summer have been deduced from satellite observations (Fishman et al. 1990). While both natural (stratospheric intrusion and lightning) and anthropogenic processes may contribute to the high levels of $O_3$, the processes need to be quantitatively evaluated (Liu et al. 1987).

Chinese atmospheric chemistry research has been conducted primarily in areas of urban pollution, for example, suspended particles, $O_3$ and $O_3$ precursors, and toxic species. Recently, there have been some important efforts to address large-scale background atmospheric chemistry issues that have regional or global implications. The major foci of these efforts include tropospheric oxidants, greenhouse gases, aerosols, stratospheric $O_3$, and acid precipitation. However, these efforts are severely limited due to a lack of funding, advanced instruments, and certain expertise in a few global change-related disciplines. It appears that atmospheric chemistry is not a field of high priority. This is also reflected in the field of atmospheric chemistry modeling, which is in its infancy compared to modeling efforts in climate or meteorology. Given the availability of highly trained theoreticians and relatively small capital investment required, one would expect to find more activities in atmospheric chemistry modeling.

The extent and nature of current atmospheric chemistry research is indicated by a survey of papers published in five leading Chinese journals[1] over a 3-year period: *China Environmental Science* (*Zhongguo Huanjing Kexue*, Chinese language, bimonthly); *Acta Scientiae Circumstantiae* (*Journal of Environmental Science* [*Huanjing Kexue Xuebao*], Chinese language, quarterly); *Environmental Science*, (*Huanjing Kexue*, Chinese language, bimonthly); *Environmental Chemistry*, (*Huanjing Huaxue*, Chinese language, bimonthly); and *Scientia Atmospherica Sinica*, (*Daqi Kexue*, Chinese language, quarterly). Between 1988 and 1990, 1,059 articles were published in these journals, and the majority dealt with water or soil pollution. A smaller number of papers dealt with atmospheric chemistry and air pollution. Out of these papers, the majority dealt with urban pollution, and only rarely with larger than regional-scale characterization. Eighty-seven of these reported results from studies of precipitation and aerosol chemistry, especially acid rain, and 25 of these papers focused primarily on aerosols.

A similar indication of the scope of atmospheric chemistry research in China is given by the proceedings of the International Conference on Global and Regional Environmental Atmospheric Chemistry held in Beijing, May 3-10, 1990. At that conference, 89 of the 173 platform or poster papers reported mainly regional and urban-scale Chinese research, and of these, 38 papers addressed acid rain. Twelve papers dealt with urban and regional oxidants and trace gases. Eight papers on aerosol studies reported measurements of chemical composition. Only eight papers discussed topics in the remote atmosphere, four of them on the stratospheric species and the other four on shipboard measurements of tropospheric trace gases and aerosols.

This record of publications indicates a preponderance of air quality and environmental impact studies that have an immediate importance to the lives of people in China. Atmospheric chemistry studies related to global climate change are far fewer. Yet, the experience gained in the environmental impact studies have laid a basis on which larger scale investigations can be conducted when future opportunities arise.

Atmospheric chemistry research is carried out primarily in a few, relatively large institutes. These include the Chinese Research Academy of Environmental Sciences (CRAES) of the National Environmental Protection Agency (NEPA), the Research Center for Eco-Environmental Sciences (RCEES) of the Chinese Academy of Sciences (CAS), the Chinese Academy of Meteorological Sciences (CAMS) of the State Meteorological Administration (SMA), CAS Institute of Atmospheric Physics, CAS University of Science and Technology of China, Peking University, and Nanjing University. Many of these institutes are involved in cooperative international research projects. However, collaboration or coordination of research activities among these institutes appears to be limited.

## Trace Gases and Oxidants

Research on trace gases other than urban air pollutants started in recent years when it was realized that all of the trace gases other than carbon dioxide ($CO_2$) contribute equally to climate change as does $CO_2$. Much of the attention has been on methane ($CH_4$), nitrous oxide ($N_2O$), and $CO_2$ emissions from various biogenic sources such as rice paddies and forests. This will be discussed below in the section that discusses atmosphere–land surface interactions.

The Chinese Waliguanshan Atmospheric Baseline Observatory for long-term monitoring of trace gases and aerosols—similar to the Global Monitoring of Climate Change stations of the U.S. National Atmospheric and Oceanographic Administration (NOAA)—is being established on the Qinghai–Tibet Plateau by CAMS with support from NOAA and the World Meteorological Organization. The plateau has an elevation of about 6 km and is remote from direct anthropogenic influence. This observatory will be open to foreign scientists when it is complete. A joint program with NOAA to measure $CO_2$ has just started. There is a plan to measure other trace gases and aerosols that play an important role in the greenhouse effect. But, shortages of funds and sophisticated instruments may delay the implementation of these measurements.

Like many cities in the world, high levels of $O_3$ are a major air pollution problem in most urban areas. While $O_3$ concentrations are routinely monitored over urban areas by local NEPA bureaux, few observations are being made in remote regions. An exception was the shipboard measurements of atmospheric $O_3$ over the western Pacific Ocean (Fu et al. 1990).

Recently, as part of International Global Atmospheric Chemistry (IGAC) projects in eastern Asia and the northern Pacific, experiments to measure concentrations of oxidants, $O_3$ precursors, $SO_2$, carbon disulfide, carbonyl sulfide, aerosols, and precipitation chemistry have been carried out separately at a number of background atmosphere stations by CAMS (in collaboration with the Georgia Institute of Technology and NOAA), Peking University, and RCEES. This work is part of the East Asia–North Pacific Regional Experiments (APARE), which include ground station measurements from China, Hong Kong, Japan, Korea, Taiwan, and the United States.

In addition, collaborating aircraft experiments over the western Pacific Ocean have been flown by the Japanese Environmental Protection Agency and the U.S. National Aeronautic and Space Administration (NASA). The experiments are called PEM–West (Chapter 4). The experiments include measurements of $O_3$, $NO_x$, nitric acid, $N_2O$, carbon monoxide, $CH_4$, non-methane hydrocarbons, $SO_2$, carbon disulfide, carbonyl disulfide, sulfate, $CO_2$, and some halogen gases. The major objectives of the experiments are (a) to evaluate the natural budgets of and anthropogenic impact on oxidants including $O_3$ and $O_3$ precursors, (b) to study the photochemical processes controlling the sulfur cycle, and (c) to study the distributions and budgets of $CO_2$, $CH_4$, and $N_2O$ over eastern Asia and the western Pacific.

The first phase of PEM–West was completed in the fall of 1991. Scientific results were presented recently at the Western Pacific Conference held in Hong Kong and will be reported in various journals and other conferences over the next few years. A second phase is being planned for the spring of 1994. The background atmosphere stations in China have made some valuable measurements. However, the capabilities of these stations are severely limited by lack of funding and advanced instruments. Without additional support, the future of these stations is uncertain.

## Aerosols

Current research in aerosol chemistry in China is more limited than in other areas of atmospheric chemistry. As discussed above, aerosol studies focus primarily on urban- and regional-scale problems. Only a few studies directly address global aerosol distributions and trends or link aerosols to climate change. From the information available to the panel, it appears that the importance of aerosols to climate change is not generally appreciated by researchers in China.

Wind-blown dust is believed to contribute significantly to particulate loading, especially in northern China (Yang et al. 1990). Dust storms carry not only soil mineral particles, but also air pollutants released from populated areas over which a dust cloud passes. These pollutants include sulfate, nitrate, soot carbon, trace metals, and organic compounds. During transport, several natural and pollution constituents may undergo chemical interaction and transformation and result in a complex aerosol mixture with atmospheric physics, chemistry, radiation, and other properties different from those of any single original soil mineral or pollution particulate constituent. Measurements of aerosols over China, Japan, and the northern Pacific have shown convincingly that dust storms originating from central Asia are the major sources of dust, sulfate, nitrate, and other particulate matter transported to the northern Pacific (Darzi and Winchester 1982, Iwasaka et al. 1988, Muayama 1988).

Given the important role played by aerosol particles in atmospheric radiation, the effect of Asian dust storms on regional—as well as global—climate needs to be carefully studied.[2] CAMS has a program to study the meteorological characteristics of dust storms, including the formation and transport of the storm's dust. A comprehensive program that addresses the chemical as well as physical properties of dust storms would be welcome. As discussed in Chapter 4, a cooperative international program called the China and America Air–Sea Experiments (CHAASE), studying the compositions of aerosols and rain over eastern Asia, has been carried out since 1990 (Arimoto et al. 1990, Gao et al. 1992a,b). As part of the Tropical Ocean and Global Atmosphere (TOGA) program, compositions of rain and aerosol samples collected over the Pacific were analyzed under a collaborative experiment between the Chinese National Research Center for Marine Environment Forecasts at the State Oceanographic Administration and NOAA.

## Stratospheric Ozone

At least four institutions, the CAS Institute of Atmospheric Physics, SMA, the CAS University of Science and Technology of China, and Peking University, are engaged in the development of one- and two-dimensional models for stratospheric chemistry studies. The CAS Institute of Atmospheric Physics and SMA have sent scientists to work with modelers in the United States.[3] Total $O_3$ is measured regularly by CAS Institute of Atmospheric Physics scientists at a station in Beijing and another in Yunnan Province. Ground-based remote sensing techniques for measuring stratospheric trace gases such as $O_3$ and nitrite ($NO_2$) are under development at Peking University, CAS Anhui Institute of Optics and Fine Mechanics, and CAMS. Of particular interest are the measurements of $O_3$ and $NO_2$ column abundances at the Chinese Great Wall Station in Antarctica (Mao 1990). NEPA has been collecting data on halon and chlorofluorocarbon (CFC) consumption over the past several years, which have been reported through the United Nations Development Program (UNDP 1992).

Because most of the stratospheric observations and laboratory measurements are carried out in the United States and Europe, Chinese modelers do not often have timely access to these data sets. Computer facilities are also somewhat inadequate to run fully coupled two-dimensional transport and chemical models efficiently. As a result, the status of stratospheric models in China is not as advanced as those in developed countries. In particular, lack of access to observational data is a serious limitation for the development of Chinese stratospheric models.

## Atmospheric Deposition

National-scale programs on the measurement of precipitation chemistry are being performed by NEPA and SMA and on the regional scale by RCEES and several provincial units. A review of these programs is difficult for two reasons. First, details of the design and structure of the programs were not available to the panel and second, the results of the programs have not been widely distributed. Furthermore, based on the panel's experience, when institutional reports are available, they often cannot be cited.

Several institutes in China have the skilled personnel to make state-of-the-art determinations of precipitation composition and indeed, have done so on a regional basis. The data sets from these studies, especially the more recent ones, are of high quality and the publications resulting from those data are of interest to the global community. However, the use of precipitation chemistry data to address global change issues requires integrated data obtained by similar methods over national scales over an extended period of time. Unfortunately, the precipitation composition data currently available are generally not adequate to address the question of China's impact on global change. Several reasons can be offered for this lack of data:

- Although several institutes measure precipitation composition on a regional basis, the collection and analytical methods are significantly different at times and so the integration of the data into a national database would be problematic.
- The quality of data from the regional programs varies with the program. Generally, the more recent data are better, but again, the absence of a standard methodology makes it difficult to come up with common quality assurance techniques.
- NEPA apparently has a database from a national monitoring network. Up to this point, it has not been possible to obtain the details of the design of the network, and it has been impossible to obtain the data. These data are not available for "outside" scrutiny.

In all fairness, it must be pointed out that the regional programs of the various institutes were not designed to address global or even national-scale phenomena. Therefore, their lack of integration, while making it difficult to address global questions, does not counter the objectives of the regional programs.

The potential for collaboration in this area is uncertain. The problems of data quality and standardization of methods are solvable, and indeed, are beginning to be addressed. The most intractable problem is one of data availability. If Chinese agencies do not provide open access to their data by scientists both within and outside China, then questions that require the use of precipitation data cannot be adequately answered.

## PHYSICAL AND ECOLOGICAL INTERACTIONS OF THE ATMOSPHERE AND LAND SURFACE[4]

### Hydrology

Because of its vital importance to agriculture and economic development, water is considered by the Chinese to be their most precious natural resource. Irrigation is at least double that of the United States and it is perhaps the most important priority in water resource policy, with hydropower ranking second and flooding ranking third. According to recent estimates, the total amount of water is about 2,800 km$^3$ (Xie and Chen 1990). The volume of water resources in China is 5 percent of the water resources of other countries in the world, and when volume is calculated on a per capita basis, China's is only 25 percent of the world average (NCCCG 1990).

The field of hydrology is a vast enterprise in China, but, as with all such activities, it is highly self-contained institutionally. However, as will be described in greater detail below, certain hydrology research programs are large enough to involve multiple institutions.

The Ministry of Water Resources (MOWR) is responsible for hydrological studies within China, and under it are seven commissions that manage the most important river systems. The Huanghe River Commission, for example, employs about 30,000 employees and responsibilities include all infrastructure—from cooking to communications. The ministry runs two training universities, the Nanjing Institute of Hydrology and Water Resources and the Wuhan Hydraulic and Electrical Engineering University, where training includes flood forecasting, water supply, and river forecasting; the work tends to be extremely practical and aimed at day-to-day operations.

In addition to MOWR, other institutions have mandates relating to water resource management. The CAS Institute of Geography is responsible for conducting surveys and research related to hydrology, mostly by using remote sensing and geographic information systems (GIS) for research tools. Other state agencies dealing with agriculture and energy are all involved in water resources one way or another. In addition, accurate forecasting of seasonal precipitation has always been SMA's function. The severe flood in the lower Yangtze River valley in 1991 further pushed precipitation forecasting up the list of priorities. Other organizations carrying out hydrology research are mainly CAMS and the CAS Institute of Atmospheric Physics.

Tang and Zhang (1989) describe research that focuses on two problems of water resource management. The first problem involves increased water pollution due to increasing population, urbanization, and demand for industrial-agricultural output. To counter this problem, hydrology research has been focused on the following aspects: (1) distribution of annual runoff and interannual variability; (2) solid discharge and its effect on the coastal regions; (3) hydrology in arid lands, urban areas, and the Huang–Huai–Hai Plain regions; and (4) lakes, glaciers, swamps, and estuaries (Tang and Zhou 1988).

The second problem concerns the uneven distribution of water resources, since most of the water supply is concentrated in the southern part of China while the northern areas have experienced increasing levels of drought in recent years. The problem is particularly severe in the major cities such as Beijing and Shanghai where groundwater supplies are almost exhausted. One strategy under investigation is to construct waterways to link the Yangtze River in the south to the Huanghe River in the north.

In 1992, the MOWR will begin a large, multi-institutional, multi-component project on the effects of climate change on hydrology. The project will continue research on the effects of climate change on basin hydrology and water resources of several major rivers located in different climate zones of the country. Subsequently, researchers will develop response strategies based on the results. For this project, the ministry wants to continue to improve research methods, for example, paying further attention to changes in the distribution of rainfall and air temperature within a given year. Four different methods are being applied to various regions:

- Statistical generation of nonstationary time series for different types of climate (pilot study area is the Haihe River)
- Large-scale hydrologic modeling on a grid and a geographical distribution model of hydrologic parameters (pilot study area is the Huaihe River basin)
- Conceptual modeling of the Yangtze and Pearl Rivers
- Statistical modeling of sediment deposition in the Huanghe River

Five tasks have been identified:

- Investigate the one-way connection between a regional climate model and a large-scale hydrologic model.
- Analyze time and space variations of precipitation and evaporation.
- Measure and analyze variations in precipitation amounts over a long series of hydrological and meteorological records.
- Research variations in the amount of water entering the sea and on the water quality of the seven major river systems.

- Research the effects of climate variations on agricultural irrigation, residential and industrial water use, electric energy production, navigation, and urban drainage, including studying adaptation strategies and conducting cost-benefit analysis.

This project will be undertaken cooperatively by the MOWR Hydrological Forecasting and Water Control Center (principal investigator: Liu Chunzheng), Nanjing Institute of Hydrology and Water Resources (Zhang Shifa), Hehai University (Liu Xinren), Wuhan Hydraulic and Electric Engineering University (Ye Shouze), the Water Science Institute of the Yellow River Basin Commission, the Hydrological Bureau of the Yangtze River Basin Commission, CAMS, and the CAS Institute of Atmospheric Physics. Other institutions may also be recruited into the project.

Relevant hydrological research at CAS includes climate change and sea-level changes, the impact of climate change on water resources, and water use efficiency studies. The project, "Preliminary Research on the Relation of China's Climate and Sea-Level Changes and its Trends and Effects," has been ongoing since 1988. It has a subproject, "Effects of Climate Change on Water Resources in North and Northwest China as well as Forecasting of Trends," that has two components—the Northwest China Project and the North China Project.

The northwest China research area includes the arid regions to the north of the Kunlun Mountains and to the west of the Helanshan Mountains, including all of Xinjiang Uighur Autonomous Region, the northern part of Qinghai Province, the Hexi Corridor in Gansu Province and the western part of Inner Mongolia Autonomous Region. Participating research units are the CAS Lanzhou Institute of Glaciology and Geocryology and the CAS Nanjing Institute of the Geography and Limnology. The Lanzhou institute is the lead unit and Shi Yafeng is the principal investigator.

The Northwest China Project has four objectives:

- Study the ice layers of the Guode ice-cap in the Qilianshan Mountains, research the ice core of the last ice age, and establish climate fluctuations over the past 10,000 years for the alpine areas in the western part of China.
- Study glacial changes of the last 500 years.
- Study changes in lake water in the arid areas in the northwest.
- Study changes of runoff of 51 mountain rivers, mountain air temperature in the summer, and annual precipitation.

The North China Project research area includes the drainage basin of the Haihe and Liaohe Rivers. Participating research units are the MOWR Hydrological Forecasting and Water Control Center, the

Nanjing Institute of Hydrology and Water Resources, the Hebei Province General Hydrology Station, and Hehai University in Nanjing. The Hydrological Forecasting and Water Control Center is the lead unit and Liu Chunzheng is the principal investigator.

The North China Project has five objectives:

- Classify, count, and statistically analyze the drought and waterlogging index data for the past 500 years and the rainfall record and air temperature of the past 100 years to ascertain spatial and temporal patterns of rainfall and air temperature in North China.
- Calculate atmospheric vapor conveyance.
- Statistically analyze the water balance variation of the land–atmosphere system by using nearly 40 years of rainfall, air temperature, and runoff records.
- Research the variations in the hydrological cycle such as precipitation and evaporation, surface water, soil water, groundwater level, storage capacity of some typical reservoirs and runoff (prompted by the fact that the 1980s were the driest 10-year period of the past 250 years).
- Calculate total annual water resources from representative drainage basins in both mountain and plains areas and study its relationship with the transformation between water and heat in the land–atmosphere system.
- Use watershed hydraulic models and climate scenarios to study the effects upon multiyear average hydrologic circumstances, such as future climate change, or annual variations of stream runoff, groundwater level, soil water, and evaporation.

This research is almost completed, and a monograph is scheduled for publication in 1993.

Water-use efficiency studies, such as those conducted at the CAS Institute of Geography and the CAS Shanghai Institute of Plant Physiology, are considered a new trend in Chinese hydrological research. CAS researchers describe a soil–plant–atmosphere continuum to refer to "sequential system studies" of the hydrological flow through that continuum (CAS 1991).

The Regional Hydrological Response to Global Warming Study Group has been approved by the executive committee of the International Geographical Union. The study group is housed at the CAS Institute of Geography (CAS 1991).

### *Summary of Hydrology*

Because the impact of drought or flood is often devastating, and because of the intense pressures on agricultural production, understanding the hydrologic cycle and the proper management of water resources are vital to China's national interests. China's approach to studying the hydrological cycle probably will remain very focused on Chinese resource management issues and on social and economic impacts. The relationship between climate change and hydrology appears to be a priority, as evidenced by the work of the National Climate Change Coordination Group and by current and planned research by major institutions such as MOWR and CAS.

The characteristics of Chinese hydrological and respective research are a very attractive area for collaboration. Under the U.S. global change research program, the Department of the Interior is conducting or planning research activities that complement the objectives of Chinese investigations. Furthermore, the U.S. Bureau of Reclamation has devised a Global Change Response Program to investigate the potential impacts of global climate change on water resources in 17 western states. Many universities, particularly in the western United States, are conducting or planning investigations of the impacts of global climate change on regional hydrology that would be conducive to collaboration with Chinese investigators.

## Biotic Controls on Trace Gases

The discipline that studies the fluxes of chemicals in the environment mediated by the biota is known as biogeochemistry. Storage of elements, especially carbon, is especially important in understanding global element cycles as changes in carbon storage can have large effects on atmospheric $CO_2$ concentrations. Biogeochemistry and physical climate can interact to produce feedbacks, if, for example, a change in climate causes an increase or decrease in a greenhouse gas such as $CO_2$ or $CH_4$, which causes a further change in climate.

The Chinese effort in biogeochemistry has numerous components relevant to land–atmosphere interactions. Here, the panel reports on several that were observed in some depth. First, a program measuring $CH_4$ emissions from rice cultivation is in progress, conducted as bilateral collaborations between CAS and the Fraunhofer Institute in Germany and between CAS and the U.S. Department of Energy (DOE). Second, the Chinese have initiated efforts to examine $N_2O$ production and $CH_4$ consumption from upland soils at RCEES. Third, an ambitious program analyzing vegetation dynamics supports national and regional estimates of carbon storage and primary productivity is under way at the CAS Institute of Botany. There are also some Chinese studies of soil nutrients and organic matter. These activities are described in more detail below.

## $CH_4$ Studies

Rice production is of considerable importance in the global $CH_4$ budget, contributing ~35-60 Tg/year out of total global emissions of 300-400 Tg. Some estimates from Chinese data suggest an even higher flux of 70-110 Tg/year (Wang et al. 1992). Rice is a major crop in China and Chinese rice production contributes a significant fraction of total global rice production, and so, presumably, $CH_4$ from rice.

Some noteworthy research on $CH_4$ from rice in China has been carried out through collaborations under the CAS-DOE Joint Research on the Greenhouse Effect (Riches et al. 1992) and with the Fraunhofer Institute in Germany (Schultz et al. 1990, Wang et al. 1992). Continuous sampling chambers allowing the collection of high temporal resolution data was undertaken in the latter collaboration. This work has made clear the significance of relatively transient high fluxes in determining overall flux. Fluxes in all Chinese studies show considerable seasonal variability and in many cases diurnal variability. In addition, the intensive nature of rice cultivation in China promotes high rice productivity and hence high $CH_4$ emissions. Different fertilizer sources did not seem to have large effects on $CH_4$ emissions integrated over the year.

Studies at the CAS Taihu Lake Comprehensive (agroecology) Experiment Station address not only $CH_4$ fluxes but also microbial and carbon cycling controls over methanogenesis. Studies are in progress to address the role of plant growth, root turnover, exudation, and microclimate and the relation to $CO_2$ emissions. Experiments are being carried out in a number of fertilizer treatments in order to understand the consequences of different agricultural management systems on $CH_4$ production.

## $N_2O$ Emissions

About 90 percent of global $N_2O$ emissions are thought to be from soils. The potential contribution of these emissions to global warming is large because it is an effective absorber of infrared radiation and very long lived in the atmosphere with a potential for thermal absorption of 150 relative to one for $CO_2$—based on turnover rate, etc. (Bouwman 1990). It is currently thought to contribute about 5 percent to radiative forcing. While the biology of $N_2O$ is relatively well known, ". . . the budget for the $N_2O$-exchange between terrestrial ecosystems and the atmosphere is largely unknown" (Bouwman 1990). The geographic distribution and magnitude of sources are poorly known. This lack of information makes additional research a priority, especially in areas of intensive agriculture and in areas where fertilizer use is increasing.

Research on $N_2O$ emissions in China is in its infancy. Currently a record of atmospheric concentrations has begun, and some flux measurements have been made. The Chinese background station (Wudaoliang) is in western China at 4,300 m elevation. Su et al. (1990), in a review paper on Chinese $N_2O$ research, report a mean atmospheric concentration of 308 ppb $+/-$ and a range from 303 to 315 ppb, data which are in the range reported from U.S. studies. $N_2O$ emissions are often linked to $CH_4$ uptake and studies on this phenomenon are beginning in China. Some work on $N_2O$ in rice paddy systems is also going on at the CAS Nanjing Institute of Soil Science.

### *Climate–Vegetation Dynamics, Net Primary Productivity, and $CO_2$–Vegetation Interactions*

The CAS Institute of Botany group led by Zhang Xinshi has an ambitious and integrated effort linking Chinese climate, vegetation, and productivity. Their effort is based on a climate–vegetation classification modified from that of Holdridge. This scheme links potential natural and agricultural vegetation to climate parameters. The effort is supported by extensive geographic data, including satellite data (advanced very high resolution radiometer [AVHRR]) for the production of a normalized difference vegetation index (NDVI). This classification, which predicts structural characteristics of vegetation, is used as a foundation for a climate-based predictor of productivity—the radiative dryness index. By using this approach, researchers have calculated net primary productivity (NPP) levels ranging from 19.5 T ha-1 y-1 on a tropical South China Sea island to <0.1 to 0.4 in extreme and temperate deserts. Because this analytical effort is fully integrated into GIS software and based on climate models, they can easily estimate national NPP levels, or calculate them under altered climates. Also, the institute has an extensive database of the chemical composition of tissues of many of the plant species of China so that budgets of nitrogen, phosphorus, and many other elements may be derived from models of NPP or biomass. This effort is the one area where modeling is central and quite strong in the field of biogeochemistry. The CAS Institute of Botany's efforts in ecosystem modeling and analysis are world-class in integration and sophistication.

Fu Congbin at the CAS Institute of Atmospheric Physics has been using NDVI for China from October 1988 through October 1989 as the basis for a pilot project on climate–vegetation interactions. The

NDVI clearly shows a vegetation boundary along the semi-arid zone of northwest China, which also delineates the northern edge of the summer monsoon. Further studies are needed to investigate the role of summer monsoons and expose interannual variability of vegetation dynamics.

$CO_2$–vegetation interactions, which has been a major research focus within the U.S. program, has received comparatively less emphasis in the Chinese research program. At least two reasons can be identified for this difference. First, facilities are limited for experimental studies of growth at elevated $CO_2$. Second, the panel did not find any perception that changes in carbon storage in terrestrial ecosystems will play a major role in mitigating or modifying $CO_2$- dependent feedback on the physical climate system. Rather, the primary Chinese interest identified to the panel is in the possible implications of $CO_2$ fertilization on agricultural yields.

One notable exception has been work at the CAS Shanghai Institute of Plant Physiology, where studies on the effects of $CO_2$ enrichment on crop growth have been conducted since 1960 (CAS 1991). Initial experiments by using elevated $CO_2$ for approximately two weeks after flowering increased rice yields by up to 30 percent and reduced the dropping of cotton balls prematurely by up to 47 percent.

In the early 1970s, experiments were conducted on cucumber, tomato, and rice seedlings that were placed under PVC tents in late spring. Cucumber yields improved by 87 percent and cucumbers reached marketable sizes 10 days earlier than those grown under control conditions. Rice seedlings grew faster and matured 3 to 4 days earlier than those in the control group.

In 1988, a $CO_2$ enrichment experiment on a natural coastal ecosystem was carried out in collaboration with D.O. Hall of London University with funding from the United Nations Environment Program (UNEP). Reed was the dominant species in the ecosystem. Two open-top PVC chambers of 2 m diameter were installed and $CO_2$ was doubled during daytime from the time new shoots appeared in the spring and continued for 44 days. Afterward, dry weight above ground biomass was increased by 44 percent and below ground by 8 percent.

Currently, a $CO_2$-doubling experiment is being carried out by Xu Daquan in Jiangsu Province, where a well containing a continuous natural flow of highly purified $CO_2$ is located. The experiment consists of four open-top chambers (two for elevated $CO_2$ and two for control purposes) containing more than ten species of weeds. This experiment will study the long-term effects of doubled $CO_2$ concentrations on plant productivity and succession.

### Soil Nutrient Studies

Soil organic matter and soil nutrient turnover has been a major research area in Western biogeochemistry (Paul and Clark 1991). Soil nutrients constrain NPP and interact strongly with climate change in controlling vegetation response. Soil organic matter is one of the largest reservoirs of carbon in the earth system and is quite dynamic in response to either land use or climate. The Chinese effort in these areas was not as evident in their global change program as were other efforts (as is the case in the United States as well) but some work is ongoing.

The CAS Nanjing Institute of Soil Science conducts considerable research on soil organic matter and has published a compendium of soil properties, including nutrients and organic matter for the major soil regions of China, as well as a national soil survey. This institute has also conducted studies of organic matter turnover in paddy soils that are integrated with $CH_4$ studies.

The CAS Northwest Institute of Soil and Water Conservation has performed extensive studies of soil and nutrient loss due to soil erosion and is involved in integrated studies of soil erosion, crop productivity, and nutrient cycling.

The CAS Inner Mongolia Grassland Ecosystem Experiment Station has published extensive results on soil carbon and nutrient pools, inorganic nutrient levels and turnover, and stable isotope biogeochemistry of grassland plants (the latter in collaboration with Larry Tieszen of Augustana College, South Dakota). Some studies on microbial ecology have also been undertaken. The work reported from Inner Mongolia and similar work conducted at the CAS Xinjiang Institute of Biology, Pedology, and Desert Research is primarily descriptive in nature but provides valuable information. The value of Chinese data on soil element storage and turnover is twofold: it augments the global database on these properties and, because soil nutrients and nutrient turnover are controls over many aspects of atmosphere–biosphere exchange, it provides context for other process studies in China, such as on trace gases or biophysics.

### Summary of Biotic Controls on Trace Gases

Biogeochemistry is a relatively young discipline. Moreover, soil nutrient turnover and trace gas biogeochemistry, while central in the global change research agenda, have been slow to develop fully, lying as they do on the disciplinary boundaries of atmospheric chemistry, soil science, microbiology, and ecology. In the United States, progress has been curbed by lack of communication between disciplines viewed as basic (for example, geochemistry or ecology) or applied (for example, soil science); these barriers exist but in a different form in China. While soil science and botany, for example, are studied under the auspices of separate institutes, they are all within CAS. Soil science and soil microbiology are central fields in global change, but as in the United States, the soil sciences are just starting to develop research agendas separate from agronomy and related to the earth and ecological sciences.

The barriers to interdisciplinary study of biogeochemistry in China are clear, given the disciplinary nature of the basic research and funding organizations. In addition, state-of-the-art research in biogeochemistry requires access to instrumentation, reliable analytical standards, and field site travel. All of these requirements can be constraining in China. However, many activities are quite vigorous and international collaboration is strong. The potential for enhanced collaboration seems high, and Chinese scientists are eager.

Two areas are especially weak in Chinese biogeochemistry. First, great progress has been made in the West by the development of techniques for the measurement of microbial processes in soil, including the conversion of organic nitrogen to mineral nitrogen, the release of organic carbon as $CO_2$ and similar phenomena for phosphorus. These techniques often rely on stable or radioactive isotopic tracers and have been fundamental in understanding processes like the rate of $N_2O$ production, the stabilization of carbon in soils and the leaching of nitrate to groundwater. These techniques require sophisticated but increasingly accessible technology for gas concentration and isotopic measurements and some sophistication in the quantitative analysis of data. The former is limiting and found at only a few institutes while the latter is abundant in China.

Secondly, in many cases, process measurements are indirect and key process rates must be estimated by system-level mass balances via modeling. While the relevant groups in Chinese institutions are aware of the significance of this type of modeling in understanding biogeochemical cycles, very little work is ongoing. Collaborative relationships could be very profitable given the mathematical sophistication of many Chinese scientists. Note also that while progress in atmospheric modeling is very limited in China by a lack of access to modern supercomputers, most ecological and biogeochemical models run satisfactorily on personal computers.

Finally, the type of teamwork and interdisciplinary collaboration that has been required for progress in biogeochemistry in the West is just developing in China. Progress will require increasingly flexible collaborative arrangements among institutions within China and encouragement of young scientists to undertake such collaborations. It will be crucial that leading senior scientists provide appropriate role models for these activities to increase in vigor. The Sino–Japanese Atmosphere–Land Surface Processes Experiment, the Huang–Huai–Hai Plain water use efficiency project, and other large-scale collaborations provide a model for Chinese research in global change in the next decade.

## Climate Change Effects on Land Cover Change Dynamics

Land cover is defined here as the vegetation and surface soil horizons or other land surfaces such as snow or ice that comprise a defined geographic area. Changes in land cover may be caused directly by human influences manifested on the global scale through climate change and enhanced $CO_2$ effects. These global causes can alter land cover indirectly as well, through pedological and geomorphological change. But, land cover change will be more probable and rapid in many parts of the world through direct intervention of human influences in land use change than through these other global factors of climate and direct $CO_2$ effects. Whereas climate change is hypothesized to occur over the next 50 years, land use change over that period is a present and continuing reality deserving attention at least equal to climate change and direct $CO_2$ effects.

Land cover change underlies some of the other focal areas addressed by this study and it is basic to some of the IGBP core projects described in this report. Land cover is a structural expression of terrestrial ecosystems underlying their contributions to energy and water budgets, biogenic trace gas fluxes, and the export of materials from land to the coastal zone through fluvial and eolian transport. For these and other reasons, land cover change is a focal area for the IGBP core project Global Change and Terrestrial Ecosystems (GCTE) (IGBP 1990).

Research on land cover change can be approached by spatial scale, by different time frames, and by different modeling methods. Several spatial scales are exhibited in the GCTE organizational plan, where land cover change will be addressed at the patch scale, the landscape scale, and the regional or continental scale (GCTE News 1991).

Different time frame approaches include past (historical analysis of change), present (monitoring of contemporary change), and future (prediction of future change). Research activities in China pertaining to land cover change are evaluated below mainly in terms of these time frames.

Historical analysis of land cover change is useful as a measure of the variance of the past, its causal factors, and its associated results. Used with caution, this information can help predict future changes. Historical analysis can be done with proxy data such as buried or altered soil profiles, preserved pollen, or tree rings; or, it may be done with direct data such as textual, photographic, or digital records that are now available from remote sensing data archives. Historical analysis is, of course, directly related to activities of the Past Global Changes Core Project (PAGES).

Monitoring contemporary land cover change provides a broad index of large-scale alterations that can be propagated to other areas such as those downstream of eolian or fluvial transport. Or, contemporary changes can be important to global processes such as land–atmosphere surface interactions pertinent to trace gas fluxes or global climate models. Monitoring can be carried out at all scales, from individual perceptions of changes in local ecosystems or crop performances, to continental scale reflectances measured by remote sensing devices.

Prediction of future change of land cover is much more difficult than monitoring contemporary change but can be approximated with a wide variety of modeling approaches such as those cited above. Prediction of change is necessary in order to evaluate the effects of policies and practices regarding greenhouse gas emissions and land use.

The predictive modeling of land cover change has no widely accepted approach. Different approaches are appropriate for different scales of concern. In general, modeling approaches can be divided into (a) correlational modeling (Box 1981), (b) patch scale/mechanistic modeling (Shugart 1984), and (c) regional scale/mechanistic modeling (Neilson et al. 1989).

## *Land Cover Change Research*

As the world's third largest country in land area, and with a long history of civilization, China has experienced climate and land use-engendered land cover change. In fact, this issue is prominent in natural resources research and in the planning of China's national global change program. Because China has a centralized form of government, the country can exert considerable influence on how the land is used (BLM 1989) and develop national land resource accounting systems and strategies for land use (Li et al. 1990). This political factor helps explain China's large amounts of data on land condition and use (Natural Resource Study Committee 1990).

*Literature.* Research on land cover and land use change—particularly on historical change—is profuse. Examples of paleobotanical treatises are Hsu (1983) and Walker (1986); change over historical time are Olsen (1987) and Richardson (1990); contemporary vegetational patterns are Tinachen (1988) and Uemura et al. (1990); the relation between climate and vegetation are Xu (1982), Xiwen and Walker (1986) and Huke and Huke (1982); changes in the soil component are Jin (1990) and Wohlke et al. (1988); the use of remote sensing are Liu (1989) and Fang et al. (1990), and on the historical transformation of the Huang-Huai-Hai Plain are Zuo and Zhang (1990).

More recently, literature has begun to emerge that treats land cover change as part of global change science, although it is no more than regional in scope. Zhu et al. (1988a) carefully define desertification as "a process of environmental degradation that is indicated by eolian activities and drifting sands caused by discordances between excessive human activities, resources, and environment on the basis of certain sand material sources under the dynamics of serious drought and frequent winds." They describe the distribution and trends in desertification for the world in general and China in particular, and analyze underlying causal factors and consequences. Most of the land defined as prone to or having undergone desertification lies in the northern tier of the country. Zhu et al. (1988b) also review these processes in more detail for a limited area of the country. More succinct analyses of desertification trends are translated from another article by Zhu and Wang (1990). Readers of broader Chinese literature should be cautioned that not all researchers use the term "desertification" as carefully and strictly as do Zhu et al. Some use the term for any form of deterioration of ecosystems or soils including the erosion of formerly forested land under moist conditions.

Chinese desertification research has a significant international link through the hosting of an international desertification training center at the CAS Lanzhou Institute of Desert Research under UNEP's auspices. Notably, in 1990, an international seminar on desertification processes was held at the Lanzhou institute (UNCSTD 1990).

Another example of global change research is in literature dealing with China's grasslands. The Gansu Grassland Ecological Research Institute (Chen 1990), in describing its national key project to monitor grasslands, states that Chinese grasslands are degrading rapidly through increase in weediness, soil degradation, and decreased productivity. This message about grassland degradation is echoed by

Reardon-Anderson and Ellis (1990). *Grasslands and Grassland Sciences in Northern China* (CSCPRC 1992) is an extensive and very sophisticated analysis of the state of change of Chinese grasslands and the underlying reasons. This study provides detailed descriptions of the degrees of conversion of grasslands to farmlands, rangeland degradation, and desertification. Collectively, the northern tier of China, especially the area surrounding marginal farmlands and grasslands appears to be relatively well studied.

Some brief discussion on "aridization tendencies" in northwest China and dryness, wetness, and chillness in east China are discussed in terms of changes in crops and forest cover in Zhu and Wang (1990). This is an example of land cover change caused, in large part, by climate change rather than from land use. In contrast with much of the literature available that focuses on historical change, a notable attempt has been made by the Second Working Team of the National Climate Change Coordination Group (NCCCG 1990) to predict the effects of climate change, at least qualitatively, on agriculture, forests, water resources, and energy.[5] Under agriculture, the report gives four examples respectively of positive and negative effects and concludes that "in short, the overall effect of climate change on agriculture will reduce China's agricultural production capacity by at least 5 percent." This seems to be a modest estimate of change compared with the descriptions of negative effects preceding the summary. After analyzing effects of climate and $CO_2$ change on six commercial species, the report offers no such summary for forests, but prognoses for those individual species seem mostly negative.

As described in Chapter 4, land cover change projections caused by climate and land use changes are attempted to a small degree, but little research was identified showing work on the direct effects of $CO_2$.

*IGBP Research.* The Chinese National Committee for the IGBP (CNCIGBP) describes a core project, "Measurement and Prediction of Changes and Trends in the Life-Supporting Environment in the next 20 to 50 Years in China," that will include land cover change, although it was not very specific in this regard (CNCIGBP 1990a). Another core project was based on setting up north–south and east–west transects of CERN stations to monitor land cover and other changes in fragile ecotones. A pilot study has been undertaken to survey existing data and Chinese literature on changes in atmosphere, vegetation cover, soils, and water.

The CNCIGBP, in 1991, announced its intention to develop five workshops to help define global change research (CNCIGBP 1991), three of which would be relevant to land cover change: (1) climate change effects on terrestrial ecosystems, (2) sensitive areas of environmental change and detection of early signals of significant global change, and (3) characteristics and trends of changes of the life-supporting environment in China.

Of the major global change research projects listed by the CNCIGBP for the Eighth 5-Year Plan, six can be easily defined as land cover change research: (1) the development and utilization of the Loess Plateau and global environmental change; (2) development and application of remote sensing techniques; (3) comprehensive land restoration experiment in the Sanjiang Plain in northeast China; (4) regional comprehensive development and land restoration strategies in southwest China; (5) comprehensive scientific study of the Karakorum-Kunlun Mountain region; and (6) study of comprehensive regional development and land restoration in Xinjiang Uighur Autonomous Region. Collectively, research on land cover change does not seem to be a distinguishable focus for the Chinese program on global change.

The Chinese Ecological Research Network (CERN), which is described in detail in Chapter 4, will be actively involved in research and monitoring of land use and land cover changes. Appendix D contains an example of the type of research planned for this network.

The National Natural Science Foundation of China (NSFC) describes projects it funds that contribute to IGBP goals (Zhang 1991). Some projects listed under PAGES might contribute to land cover change research from an historical viewpoint. Some projects listed under the "Human Dimensions of Global Change Program" look very interesting for further modeling of the land use change component of land cover change. See Appendix B.

A jointly sponsored (U.S. Environmental Protection Agency, NEPA, CAS) symposium on climate–biosphere interactions was held in Beijing in May 1991, during which participants were asked to recommend future collaborative work. Concerning models, participants recommended four areas for cooperation: (1) data compilations on land cover and use in Asia including inputs to large-scale models; (2) specific emphasis on rates of reforestation/afforestation; (3) cooperative AVHRR land cover study as part of the Data and Information Systems for the IGBP and inclusion of other satellite data; and (4) regional, mesoscale, forest, agricultural, and soil ecosystem models with processes.

Research activities that could broadly be interpreted as addressing land cover change can be found in many of the institutional reports in Appendix A; the following give a sampling of the diversity of activities.

The Commission for Integrated Survey of Natural Resources (CISNAR) has a role to play in land use and land cover change research. The many expeditionary and cartographic activities it carries out would be an excellent source for the development of a contemporary land cover database for the nation. Institute officials say it will be the database center for land resources in the national global change program. CISNAR already houses World Data Center-D for Renewable Resources and Environment and the Integrated Research Center for Natural Resources and Agricultural Development (a joint database project with the Ministry of Agriculture) and has regional and national databases for resources, ecology, and the environment. Additionally, CISNAR is the designated host of the CERN Synthesis Center, where synthesis and analysis of CERN-generated data will be conducted. Clearly, CISNAR has an important role to play and it is hoped that it will receive the extensive training, expertise, and appropriate equipment that will be required to realize this role successfully.

Through CISNAR, CAS has produced a very impressive compendium of natural resources that indicates a kind of database management (Natural Resources Study Committee 1990). A little historical analysis is visible here in the way of dendrochronology, but no evidence of predictive modeling activities.

Feng Zongwei and Zhuang Yahui at RCEES are carrying out a large-scale, collaborative forest ecology project with the objective of providing specific data on China's forest ecology for input to global change studies. Methodology includes soils and plant classification and mapping, experimental studies of material and energy flows, measurements of atmospheric gas concentrations ($CO_2$, $NO_x$, $SO_2$, and $O_3$), and comparison of differences between urban Beijing atmosphere and nearby mountain atmosphere. Comparisons will be made in different seasons, trends will be monitored for 5 years, and some modeling will be included. The mountain area studies will be conducted at the CAS Beijing Forest Ecosystem Station, which is located about 2 hours west of Beijing. The cost of the project is 1.8 million *yuan*, and funding is from the State Science and Technology Commission and NSFC.

In Nanjing, a few institutes are or could be contributing to land cover change research. At the CAS Nanjing Institute of Geography and Limnology, considerable historical analysis could be relevant. At the Department of Geo and Ocean Sciences at Nanjing University, high-quality historical analysis and qualitative land cover change projections are being developed. The projections are done mainly through land use change, and are coupled with hydrological and sedimentological consequences. (One of the most holistic and imaginative proposals on these linkages has been proposed by Ren Mei-e to NSFC, and a preliminary manuscript [Ren and Zhu 1991] already shows some very interesting results.) Under NEPA, the Nanjing Institute of Environmental Science is conducting work on preserving biological diversity, deforestation and reforestation, and desertification, although little evidence exists of any kind of spatially extensive program or aspiration to such an approach.

The CAS Institute of Botany presented the most sophisticated example of predictive capability that members of this panel encountered. Potential natural vegetation defined by the Holdridge system of physiognomic–climate relationships and agricultural cropland are linked with climate variables through discriminate analysis (Chang and Yang 1991). Climate variables are interpreted in terms of large-scale drivers for the subcontinental weather system such as the Qinghai–Tibet Plateau and East Asian monsoon and therefore are linked to potential scenarios of large-scale meteorological expressions of climate change.

This entire system is built on national-scale GIS with variable grid cell sizes for different attributes. The result is a ready system for portrayal and manipulation of vegetation data for the country. The nationwide soils map is also in this system. This is a prime example of the application of a simple correlation modeling approach to large-scale prediction of vegetation change that has no readily apparent comparison elsewhere in the world. Other, more sophisticated, mechanistic models exist but are not applied to a large-scale database system. With more sophisticated models, first order assessments of direct $CO_2$ effects could be made when more physiological understanding exists.

This GIS system is assessed with satellite imagery country-wide and is in a format that can be used for equilibrium prediction for altered climates. Predicted changes in annual precipitation or temperature are used to recalculate productivity and regenerate maps. In this system, China has a tool for assessing the impacts of climate change and land use change on large-scale land cover and for evaluating the climate, hydrological, and ecological consequences through linked models. These are not dynamic models in the sense that they cannot simulate *rates* of change. They are, however, comparable in sophistication to similar models employed in the West.

The CAS Institute of Geography's historical and contemporary land cover studies are highly germane to the study of land cover change. The National Key Laboratory for Resources and Environment Information Systems conducts some of the most comprehensive and sophisticated GIS work in China, although it is notable that it is done in complete isolation from other solid work such as that described above at the Institute of Botany. At a recent international meeting on GIS sponsored by the laboratory, 17 of the 55 Chinese papers submitted were generated by the CAS Institute of Geography. Seven papers each were presented from Wuhan and Peking Universities, four (mainly statements of need) from CISNAR, three from Nanjing University, two from the CAS University of Science and Technology of China, and one each from 13 other organizations (Chen Shupeng et al. 1990). Although a marked contrast in the technical sophistication of these papers was evident, this brief survey was remarkable and encouraging, nevertheless, as to the potential for developing GIS and remote sensing capacities in China.

*Summary of Land Cover Change Dynamics*

In order to describe land cover change in the past and present, or to make predictions, appropriate land cover classifications have to be devised and broadly accepted at appropriate scales, and there has to be sound geographic control of land cover in the form of maps or digital database systems. Although a methodological review of map resources was not conducted, it would seem that China is rich in fine soil and vegetation maps at local to national scales. A quick review of information available through CAS shows many valuable components that could contribute to a more systemized land cover accounting system. For example, the National Key Laboratory of Resources and Environment Information Systems at the CAS Institute of Geography has substantive mapping and GIS capabilities. China receives Landsat data and has the resources of the CAS Institute for Remote Sensing Applications. And, digitizing maps is a project within CERN. Ultimately, an appropriate GIS system and coordinated remote sensing system for the country will be needed to bring order to the spatial data and to monitor continuing land cover change expediently.

The other ingredient needed for projecting land cover change is a system of relating vegetation components (species or functional groups) to climate for natural vegetation, cropping systems to climate, $CO_2$ increase, and social and economic factors through the use of models, such as the modeling objectives at the CAS Institute of Botany described previously. Also, current plans for the CERN Synthesis Center at CISNAR call for collecting such data and for having the resources to build those types of models.

Contemporary and planned research on land cover change in China is fragmented and largely historical in approach. Research currently is a national program on phenomena driven, in part, by global mechanisms. Within China, there seems to be more progress in the north, particularly on lands surrounding and including grasslands, than in the far northeast or south. Much of the work under way is in the category of historical analysis.

China does not appear to have a comprehensive or interinstitutionally systematic monitoring system. CAS has launched a major program to create a network of ecological stations that will have monitoring functions (Chapter 4). SMA has over a thousand monitoring stations. NEPA has hundreds of various types of monitoring stations and is implementing a national monitoring system. Various ministries, such as the Ministry of Agriculture, monitor resources under their jurisdiction.

It would be useful to nurture modeling efforts that use more mechanistic approaches to modeling the relationships between vegetation and soils (as the elements of land cover) and climate and land use. The panel did not find that such activity is planned, even though this is key for global change research. The obvious importance to policy makers will lie in having predictive descriptions of land cover change.

In China, extensive land cover changes have already occurred and are ongoing today. Much of this is driven by land use change rather than by changing climate *per se*. Many components are available to develop a focused land cover change program in China: maps, remote sensing, historical records, and some modeling capacity. This is an area in which some catalytic action through international collaboration could make a big difference.

Additionally, as in other countries, an entirely different genre of social science models is needed for predicting how land use would be likely to change in response to climate, population, economic, and technological change scenarios. Furthermore, land cover change research must have strong ties to research on the human dimensions of global change, which, in the case of China, is an area of great relevance and potential.

## NOTES

1. In China, it is common for individual institutes to publish journals of the institute's research work, often without external peer review. It is notable, though, that some journals are now accepting substantial numbers of papers from outside institutes.

2. In June 1992, a workshop on Asian dust was organized by Richard Arimoto, University of Rhode Island, that brought together more than 30 scientists from China, Japan, and the United States to report on recent research and to discuss future cooperative studies.

3. Under a bilateral project between NASA and CAMS, two workshops on atmospheric chemistry were held in 1990 and 1992, and stratospheric $O_3$ was one of the selected topics. Each workshop involved between six and 15 scientists from each country who presented key findings from their research.

4. This area has been nourished by international exchanges and by the strong scientific leadership of Ye Duzheng, who has collaborated extensively with foreign and Chinese research communities, bringing them together. Early on, Ye emphasized the important and unique role of the Qinghai–Tibet Plateau as a strong dynamic and thermodynamic perturbation on the global general circulation. He organized measurement programs and data collection to examine this effect and he has carried out a numerical sensitivity study of the effects of soil moisture on climate and hydrological variability.

5. This report was drafted for the Intergovernmental Panel on Climate Change's deliberations.

# 6

# Summary

## INTRODUCTION

The major thrust of this report has been to ask—in a systematic way—how Chinese science is responding to global change and to identify opportunities for and constraints to collaboration in global change research. One of the guiding beliefs of the panel has been that successful collaboration stems from an understanding of respective capabilities, respect for national priorities, and a desire to share knowledge and information. In this light, the report has emphasized opportunities for global change science for the Chinese themselves, for potential collaboration with the U.S. scientific community, and for the integration of Chinese research into major international global change research programs.

Even though China is a developing country, it has a substantial scientific infrastructure. As this report has demonstrated, China is organizing a global change program of significant potential. As such, it is well positioned among developing countries to respond to calls from the international scientific community to participate in global research regimes. The extent of Chinese participation in collaborative global change projects will be determined by China's domestic priorities, availability of funding, human resources, scientific capabilities, and the responsiveness of the international scientific community to these factors.

In order to understand opportunities for global change research in China, it is useful to look at the panel's findings from both top-down and bottom-up perspectives. Science in China is driven by government policies and priorities, and so is inherently a top-down enterprise. First, governmental interest and resources definitely favor studies of the impacts of environmental change on China, as opposed to the impact of China on global change. Consequently, Chinese researchers have focused on questions that stress the human dimensions of environmental change and on the social and economic impacts of climate change. Second, funding for global change research has been slow in developing. The main source for these funds, the National Natural Science Foundation of China (NSFC), does not have the budget necessary to fund the broader based or larger scale research that many global change issues require. Third, funding and research are organized along disciplinary lines, which complicate the development of interdisciplinary research. Fourth, scientific institutions are vertically organized and integrated, which can discourage cooperation.

From the bottom-up perspective, the Chinese scientific community almost universally considers working on global change topics to be a challenge, an opportunity and, given China's vulnerability to climatic disasters, a responsibility. Moreover, these scientists are pursuing this research with energy, commitment, and creativity. It is important to note that many of the problems Chinese scientists face in developing a global change research program are similar to those faced by scientists in the United States, including resistance from institutions organized to address traditionally well-defined and more disciplinary problems, lack of appropriately cross-trained scientists, and the difficulties of developing funding for new programs.

Problems of data availability and the cost of data indicate the lack of organizational integration and policies that promote proprietary interests. When data are not available across structures, institutions respond by generating the data they need themselves. When institutions can control data, then data can become commodities. Moreover, decreases in funding are increasing the pressure to sell data profitably.

The way Chinese data are managed can be a major stumbling block to collaboration. Resources and rewards are few for carrying out basic documentation. Because data sharing and intercomparison are not promoted, the incentive for documenting data for others' use is reduced. Because research results are often published in journals of the institutes in which the research was conducted, the lack of external peer review by the broader scientific community reduces the motivation to document one's work. Appropriate documentation of data and procedures does not appear to be a regular criteria in publishing.

## CONTRIBUTIONS TO INTERNATIONAL RESEARCH PROGRAMS

It is important to emphasize that the panel did not set out to provide an exhaustive investigation into all aspects of science and policies relevant to global change. Rather, by using Chinese national committees to indicate the organization of global change research, the panel provided an overview of activities relevant to the International Geosphere–Biosphere Program (IGBP) and the World Climate Research Program (WCRP). More in-depth reporting was confined to specific areas of interest to the panel. Consequently, the panel's findings are not comprehensive, nor are recommendations made that would overreach the information the panel was able to collect.

Ideally, a successful global change research program would consist of fully integrated institutions, staffed with well-trained and enthusiastic scientists who work with sophisticated equipment, generate high-quality databases, have a robust modeling capability, and enjoy strong government support on the questions pertinent to the issue. In the panel's opinion, no global change research program in the world meets these criteria. Thus, while problems exist in the Chinese global change program and in the way science is organized in China, it should be pointed out that the Chinese are in good company. More importantly, the Chinese are already making significant contributions to global change research and have the potential for even greater contributions. In the context of international collaboration, these considerations could be even more significant if the following findings are considered:

- Increased research on the impacts of China's pollution, especially on coal combustion emissions, when coupled with assessments of these impacts on the global atmosphere, oceans, and terrestrial ecosystems, would be a contribution to international understanding of global change processes.
- Because science is closely tied to economic considerations, China has the potential to make significant contributions to research on the human dimensions of global change.
- China's geography offers important and unique opportunities for global change studies. For paleoclimate research, the Loess Plateau has accessible loess deposits containing an extensive

paleorecord. The Qinghai–Tibet Plateau itself modifies climate as well as being a site for paleoclimate research on its midlatitude glaciers and saline lakes. China's extensive and highly populated coastline will provide opportunities to study land–ocean interactions in the coastal zone.
- China's extensive historical writings contain proxy data for climate change that are a unique contribution to research on past changes.
- Historical analyses of climate, land cover and land use, hydrology, sediment flux, and sea-level changes constitute an extensive component of Chinese research. Contributions will be strengthened by coupling these analyses with prospective research designed to provide projections of future changes.
- Ministries play key roles in disseminating emissions data that are crucial inputs to global models. Data currently available are not sufficient to carry out research that is important to Chinese policy makers as well as the international community.
- The establishment and availability of national inventories of fossil fuel emissions, trace gas production rates, and soil properties (e.g., texture, organic carbon storage, and water-holding capacity) will strengthen Chinese contributions.
- Increased funding through NSFC would expand opportunities for interdisciplinary and collaborative global change research.
- Research on surface and radiative fluxes in agriculture and observational programs in hydrology are relevant to global change research programs. Contributions from agricultural and water resource research institutions are expected to increase as they become more fully integrated under the rubric of global change studies.
- Interest in developing global change education programs to meet both national and regional needs deserves further attention and funding.
- The biosciences are poised to make substantive progress if more resources such as computers and training in modeling were available. Increasing access to computers would also force improvements in data management, including documentation and quality assurance and quality control procedures.
- Agreements concerning access to, standards, and sound management of data would strengthen Chinese science itself as well as encourage cooperative research.
- Contributions will be determined in part by progress in reforming science education and in attracting students back to China after they receive Ph.Ds in foreign countries.

SUMMARY                                                                  123

## PROSPECTS FOR COLLABORATION

For Chinese science, cooperative projects have been a vehicle for the transfer of knowledge, techniques, and equipment and have created important training opportunities. Domestic funding for research remains very limited, and governmental support will likely remain driven by domestic priorities. The level of collaboration and participation in international global change research programs will be highly dependent upon funding from international sources.

Interest in carrying out cooperative science in China is driven by scientific questions. On the Chinese side, this mutual interest is additionally driven by training, equipment, and funding needs. Consequently, reaching agreement on project details can be time consuming and complex. The panel found a high potential for increased cooperation in all areas it examined. But, the development of actual projects will require substantive resources to organize and administer the projects. Furthermore, it will be important to fund projects that allow foreign scientists to collaborate in China for longer periods of time, such as from 6 months to 1 or more years. These conditions will be especially important in larger-scale projects and in any project requiring access to tightly controlled or fragmented sources of data.

Prospects for collaboration with American scientists have been adversely affected by increasing political tensions between the United States and China that began with the Chinese government's crackdown on students in Tiananmen Square in 1989. Since then, the slowing American economy and tightened research budgets have caused a retrenchment and rethinking of American science priorities. Still, as this report attests, American and Chinese scientific cooperation has been built on a strong foundation of mutual interest and respect, which will continue to foster cooperative initiatives even in leaner times.

The panel was not charged with making comprehensive recommendations about how the U.S. government or the organizing units of the international research programs should address China's role in their respective global change research programs. Rather, the panel viewed its charge mainly as an exercise in organizing information about Chinese global change research in order to increase foreign understanding, which would, in turn, stimulate initiatives with Chinese colleagues at many levels of interaction. To this end, the panel makes the following suggestions for enhancing collaboration:

- Expanded and diversified Chinese participation in international global change scientific meetings, particularly in modeling, training, and data and information, would enhance the Chinese glo-

bal change research program and increase opportunities to initiate collaboration.
- Global change is a current priority of the Pacific Science Association, which could be an organizing force for increasing China's participation in international global change research.
- Collaboration on paleoclimate would provide foreign access to these valuable and often unique data and would facilitate the addition of predictive capacities to existing Chinese strengths in descriptive paleoresearch. Collaboration would facilitate the consideration of relationships with the broader global system that may have driven some of these past phenomena.
- Establishing policies that promote the availability of data and that limit the cost of data acquisition to reasonable cost recovery will make collaboration more attractive to foreign scientists.

# References

Abramson, M. 1990. Open labs in China. China Exchange News, (March) 18(1):8–13.
Arimoto, R., Y. Gao, R.A. Duce, D.S. Lee, and L.Q. Chen. 1990. Links between the land, atmosphere and oceans: The biogeochemical cycles of trace elements. Pp. 24–40 in Global and Regional Environmental Atmospheric Chemistry. Proceedings of the International Conference on Global and Regional Environmental Atmospheric Chemistry, L. Newman, W. Wang, and C.S. Kiang, eds. Springfield, Va: National Technical Information Service.
Bouwman, A.F., ed. 1990. Soils and the Greenhouse Effect. Chichester: John Wiley & Sons.
Box, E.O. 1981. Macroclimate and Plant Forms: An Introduction to Predictive Modeling in Phytogeography. The Hague: Dr. W. Junk Publishers. 258 pp.
Bradley, R.S., ed. 1991. Global Changes of the Past. Papers arising from the 1989 Global Change Institute, Snowmass, Colorado, 24 July - 4 August 1989. Boulder: University Corporation for Atmospheric Research/Office for Interdisciplinary Earth Studies. 514 pp.
Bureau of Land Management (BLM). 1989. Report on the State Land Administration in the People's Republic of China. Washington D.C.: U.S. Department of the Interior.
Chinese Academy of Sciences (CAS). 1978. Handbook of Plants of Dinghushan [Arboretum]. 647 pp. (In Chinese with names in Latin.)
CAS. 1991. Global Change Study in Chinese Academy of Sciences. (Multiple page sets, in English.)

CAS. Undated, a. The Chinese Academy of Sciences. Beijing: Chinese Academy of Sciences. (In English.)

CAS. Undated, b. Vegetation Map of Dinghushan Biosphere Reserve. 2 pp. (In Chinese with legend in English.)

Chang Hsin-shih (Zhang Xinshi) and Yang Dianan. 1991. A Study of Climate-Vegetation Interaction in China—The Ecological Model for Global Change. Preprint. Beijing: Institute of Botany. 29 pp. (In English.)

Chen Jiaqi. 1986. The heavy flood and drought of Taihu Basin from the Southern Song Dynasty and the possibility of their reoccurrence in the near future. Scientia Geographica Sinica 6(1). (In Chinese with abstract in English.)

Chen Jiaqi. 1987. An approach to the data processing of historical climate materials on the basis of floods and droughts of the Taihu Basin. Acta Geographica Sinica 42(3). (In Chinese with abstract in English.)

Chen Jiaqi. 1989. A preliminary research on the regularity and factors of changes of flood and drought in Taihu Basin since the southern Song Dynasty. Scientia Geographica Sinica 9(1). (In Chinese with abstract in English.)

Chen Jiaqi. 1991. The law and tendency of variation of flood/drought in the last 1,500 years along the banks of the middle Huanghe River. 2 pp. (In Chinese with abstract in English.)

Chen Jiaqi. Undated, a. Preliminary study on regularity of climatic variation in the Taihu Lake Region. (In Chinese with abstract in English.)

Chen Jiaqi. Undated, b. The great drought and flood (sic.) in the Huanghe River Basin during historical time. Memoirs of Nanjing Institute of Geography, Academia Sinica No. 1. (In Chinese with abstract in English.)

Chen Q. 1990. The study of developing the grassland resources remote sensing monitoring system in China. Gansu Grassland Ecological Research Institute, Lanzhou, China. (Unpublished project description in English.)

Chen Shixun. 1990. Summary Report of Climate Record for Past 2,000 Years in Guandong Province. 106 pp. (In Chinese.)

Chen Shupeng, He Jianbang, Zhao Chunyan, Zhong Ershun, Yuan Xiansheng, and Hu Jue, eds. 1990. Proceedings of the Second International Workshop on Geographical Information Systems, 8–11 August 1990, Beijing. Beijing: National Laboratory of Resources and Environment Information Systems. (In English.)

Chinese Joint Global Ocean Flux Study Committee. 1989. Chinese Tentative Plan for the Joint Ocean Flux Study (JGOFS). 8 pp. (In English.)

Chinese National Climate Committee. 1990. Outline of National Climate Program of China (1991–2000). Beijing: China Meteorological Press. (In Chinese and English.)

Chinese National Committee for the IGBP (CNCIGBP). 1990a. Status Report of the Chinese National Global Change Research Program. A report to the first meeting of the national committees, Washington, D.C., January. 13 pp. (In English.)

CNCIGBP. 1990b. A report to the Second Scientific Advisory Council of the

International Geosphere–Biosphere Program. September. 9 pp. (In English.)
CNCIGBP. 1991. Bulletin of CNCIGBP, (May) 1(1). Reprinted in Global Change Newsletter, December 8, 1991. (In English.)
CNCIGBP. 1992. Proposal to IGBP START Standing Committee: To Establish a Global Change Regional Research Network for East Asia and Western Pacific Region. Beijing, March 12, 1991. (Draft proposal in English.)
Committee on Scholarly Communication with the People's Republic of China. 1992. Grasslands and Grassland Science in Northern China. Washington D.C.: National Academy Press.
Darzi, M. and J.W. Winchester. 1982. Aerosol characteristics at Mauna Loa Observatory, Hawaii after East Asian dust storm episodes. Journal of Geophysical Research 87:1251–1258.
Fang Liping, Li Qi, and Cheng Jicheng. 1990. Remote sensing dynamic monitoring method of land-use/land cover. Pp. 91–100 in Proceedings of the Second International Workshop on Geographical Information Systems. Chen Shupeng, He Jianbang, Zhao Chunyan, Zhong Ershun, Yuan Xiansheng and Hu Jue, eds. Beijing: National Laboratory of Resources and Environment Information Systems.
Fishman, J., C.E. Watson, J.C. Larson, and J.A. Logan. 1990. Distribution of tropospheric ozone determined from satellite data. Journal of Geophysical Research 95:3599–3618.
Fu Jimeng, Zheng Xianying, Wen Yupu, and Su Weihan. 1990. Preliminary shipboard measurements of atmospheric ozone over the western Pacific Ocean. Pp. 180–183 in Global and Regional Environmental Atmospheric Chemistry. Proceedings of the International Conference on Global and Regional Environmental Atmospheric Chemistry. L. Newman, W. Wang, and C.S. Kiang, eds. Springfield, Va: National Technical Information Service.
Gao Y., R. Arimoto, M.Y. Zhou, J.T. Merrill, and R.A. Duce. 1992a. Relationships between the dust concentrations over eastern Asia and the remote North Pacific. Journal of Geophysical Research 97:9867–9872.
Gao Y., R. Arimoto, R.A. Duce, D.S. Lee, and M.Y. Zhou. 1992b. Input of atmospheric trace elements and mineral matter to the Yellow Sea during the spring of a low dust year. Journal of Geophysical Research 97:3767–3777.
GCTE News. 1991. Newsletter of the global change and terrestrial ecosystems (GCTE) core project of the IGBP. (August) 1:1–12.
Gosz, J., B. Leach. 1992. U.S.-China exchange laying the foundations for collaboration. LTER Network News 10:1.
Hsu J. 1983. Late Cretaceous and Cenozoic vegetation in China emphasizing their connections with North America. Annals of the Missouri Botanical Garden 70:490–508.
Huke, R.E. and E.H. Huke. 1982. Agroclimatic and dry-season maps of South, Southeast, and East Asia. Los Baños: International Rice Research Institute. 15 pp.

International Council of Scientific Unions (ICSU). 1992. International Council of Scientific Unions Year Book 1992. Paris: ICSU.

International Geosphere–Biosphere Program (IGBP). 1990. International Geosphere–Biosphere Program: A Study of Global Change, the Initial Core Projects. IGBP Report No. 12. Stockholm: IGBP.

IGBP. 1991. IGBP Report No. 15. Global Change System for Analysis, Research, and Training (START). Boulder: UCAR Office for Interdisciplinary Earth Studies.

IGBP. 1992a. A Quick Start for START, Guidelines for Regional Research Networks and Centers. Global Change Newsletter, 9 (March).

IGBP. 1992b. START from Concept to Reality. Global Change Newsletter 10 (June).

Intergovernmental Panel on Climate Change (IPCC). 1990. Climate Change, the IPCC Assessment, J.T. Houghton, G.J. Jenkins, and J.J. Ephraums, eds. New York: Cambridge University Press.

Iwasaka, Y., M. Yamato, R. Imasu, and A. Ono. 1988. Transport of Asian dust (kosa) particles: Importance of weak kosa events on the geochemical cycle of soil particles. Tellus 40b:494–503.

Jin Zhengping. 1990. Investigation of the present status of soil erosion conditions with remote sensing technology in Inner Mongolia. Soil and Water Conservation in China 2:26–28. (In Chinese and English.)

Khalil, M.A.K., R.A. Rasmussen, M.X. Wang, and L. Ren. 1990. Emissions of trace gases from Chinese rice fields and biogas generators: $CH_4$, $N_2$, CO, $CO_2$, chlorocarbons, and hydrocarbons. Chemosphere 20:207–226.

Leach, B. 1990. Long-term ecological research in China: CAS establishes a network. China Exchange News 18(4):23–27.

Leach, B., J. Brunt, W. Michener, J. Gosz, K. Gross, J. Pastor, D. Coffin, A. McKee, C. Bowser, S. Batvik, and A. Mossige. 1992. A Report on the Chinese Ecological Research Network (CERN). Submitted to the World Bank. February.

Li Debo, Zhang Jiwu, Li Weixing, and Yu Fei. 1991. Monitoring and controlling of $CH_4$ emissions from rice paddy field near Nanjing, China. 13 pp. (Preprint in English.)

Li G.L. 1990. The National Natural Science Foundation of China: Its background, present, and guiding principles. Science and Public Policy 17(4):235–241.

Li Jinchang, Gao Zhengang, Zhang Zhaoxiu, He Xianjie, Kong Fanwen, and Fu Dedi. 1990. Natural Resources Accounting For Sustainable Development. Development Research Center of the State Council. Beijing: China Environmental Science Press. 259 pp.

Liu Leyi. 1989. Surveying the present state of utilization of land resources by using remote sensing technology. Soil and Water Conservation in China 1:21–23. (In Chinese and English.)

Liu S.C., M. Trainer, F.C. Fehsenfeld, D.D. Parrish, E.J. Williams, D.W. Fahey, G. Hubler, and P.C. Murphy. 1987. Ozone production in the rural troposphere and the implications for regional and global ozone distributions. Journal of Geophysical Research 92:4191–4207.

Liu Tungsheng, ed. 1991. Loess, Environment and Global Change. Beijing: Science Press. (In English.)
Ma Fuchen. 1991. One-page summary of national thrusts relevant to PAGES. (In English.)
Mao J. 1990. Measurement of $O_3$ and $NO_2$ column abundance at the Great Wall Station, Antarctica. Pp. 180–183 in Global and Regional Environmental Atmospheric Chemistry. Proceedings of the International Conference on Global and Regional Environmental Atmospheric Chemistry, L. Newman, W. Wang, and C.S. Kiang, eds. Springfield, Va: National Technical Information Service.
Muayama, N. 1988. Dust clouds "kosa" from Asia dust storms in 1982–1988 as observed by the GMS satellite. Meteorological Satellite Center Technical Note No. 17.
National Climate Change Coordination Group (NCCCG). 1990. Impact of Human Activities on Climate in China. *Zhongguo Huanjing Kexue* (Chinese Environmental Science). 10(6) December (In Chinese.) English text in JPRS-TEN-91-010, 11 June 91.
National Natural Science Foundation of China (NSFC). 1990. Guide to Programs, Fiscal Year 1990. Beijing: Science Press. (In English.)
NSFC. 1991a. Briefing paper concerning support of JGOFS in the PRC. 1 p. (In English.)
NSFC. 1991b. Briefing paper concerning support of WOCE in the PRC. 1 p. (In English.)
National Science Foundation (NSF). 1991. Arctic Science, Engineering, and Education, Directory of Awards, Fiscal Year 1991. Washington, D.C.: NSF.
Natural Resource Study Committee. 1990. Handbook of Natural Resources of China. Beijing: Chinese Academy of Sciences, 902 pp. (In Chinese.)
Neilson, R.P., G.A. King, R.L. DeVelice, J. Lenihan, D. Marks, J. Dolph, W.G. Campbell, and G. Glick. 1989. Sensitivity of Ecological Landscapes and Regions to Global Climatic Change. Washington, D.C.: National Technical Information Service.
Nie Baofu, Liang Meitao, Zhu Yuanzhi, Zhong Jinliang, and Hua Ouyang. 1991. The study of the fine skeletal structures of recent reef-corals in the reef-coral regions of the South China Sea. Guanghzou: CAS South China Sea Institute of Oceanology, 148 pp. (In Chinese with English abstract.)
Nie Baofu. Undated. Approach to the relationship between growth rate of some reef corals and surface water temperature in central and northern parts of the South China Sea. Unknown journal reprint. (In Chinese with English abstract.)
Olsen, J.W. 1987. Prehistoric land use and desertification in northwest China. MASCA Journal 4:103–109.
Paul, E.A. and F.E. Clark. 1991. Soil Microbiology and Biochemistry. San Diego: Academic Press.
Pu Peimin. 1983. Studies on the influence of tide-generating forces of celestial bodies on weather–climate from the factors of some quasi-period climatic oscillations. Unknown outlet. (In Chinese with English abstract.)

Pu Peimin. Undated. Research activities on climate change and its impact on water resources in China. 5 pp. (In Chinese and English.)
Qian Junlong. Undated, a. A study on dynamic change of environments from the contents of chemical elements in tree rings. 6 pp. (Unpublished summary of work in English.)
Qian Junlong. Undated, b. A study on relationships between the content of chemical elements in tree rings and the related environment. Photocopy of reprint with no source given. (In English.)
Qian Junlong. Undated, c. Determination of 22 elements in growth rings of trees by ICP-AES. 6 pp. (Unpublished summary in English.)
Research Center for Eco-Environmental Sciences (RCEES). 1991. 1990 Annual Report. Beijing: RCEES. (In Chinese with abstract in English.)
Reardon-Anderson, J., and J. Ellis. 1990. Whither China's grasslands? China Exchange News 18(4):13–17.
Reed, Linda A. 1989. Education in the People's Republic of China and U.S.-China Educational Exchanges. Washington, D.C.: NAFSA.
Ren Mei-e and Zhu Xianmo. 1991. Anthropogenic effect on changes of sediment discharge of the Yellow River, China since the Holocene—A preliminary study. Quaternary Studies 1.
Ren, Mei-e. 1991a. Relative sea-level rise in Huanghe River delta, China—Implications and response strategies. Preprint for the Proceedings for the Social Impact of Sea-Level Rise, E. Stoddard, ed. Berkeley: University of California Press. 6 pp. (Preprint in English.)
Ren Mei-e. 1991b. Anthropogenic effect on changes of sediment discharge of the Huanghe River in China Since the Holocene: A Preliminary Study. 3 pp. (Proposal outline in English.)
Richardson, S.D. 1990. Forests and Forestry in China: Changing Patterns in Resource Development. Washington, D.C.: Island Press. 352 pp.
Riches, M.R., Zhao J.P., W.C. Wang, and Tao S.Y. 1992. The U.S. Department of Energy and the People's Republic of China's Academy of Sciences joint research on the greenhouse effect: 1985–1991 research progress. Bulletin of the American Meteorological Society 73(5):585–594.
Saich, T. 1989. China's Science Policy in the 80s. Atlantic Highlands, N.J.: Humanities Press International.
Schultz, H., W. Seiler, and H. Rennenberg. 1990. Soil and land use related sources and sinks of methane ($CH_4$) in the context of the global methane budget. Pp. 269–285 in Soils and the Greenhouse Effect, A. F. Bouwman, ed. Chichester: John Wiley & Sons.
Shi Yafeng. 1991. Studies on climatic and sea level changes in China funded 1987–1992 by NSFC and CAS. 2 pp. (Project description in English.)
Shi Yafeng, Wang Mingxing, Zhang Peiyuan, and Zhao Xitao, eds. 1990. Studies on Climatic and Sea Level Changes in China I. Beijing: China Ocean Press. 146 pp.
Shugart, H.H. 1984. A Theory of Forest Dynamics: The Ecological Implications of Forest Dynamics Models. New York: Springer-Verlag. 278 pp.
State Science Technology Commission (SSTC). 1990. Climate (abridged),

No. 5 Blue Book of Science and Technology. Beijing: State Meteorological Press. (In Chinese with abridged version in English.)
State Planning Commission. 1991. National Report of the People's Republic of China on Environment and Development. Beijing: SPC. 83 pp. (In Chinese and English.)
Su Weihan, Zeng Jianghai, Song Wenzhi, Zhang Hua, Zhang Yuming, Cao Meiqiu, Lu Hongrong, and Zhou Quan. 1990. Flux of nitrous oxide on typical winter wheat field in northern China. Photocopy. (Unpublished manuscript in Chinese.)
Tang Q. and Zhou C. 1988. Recent progress of regional hydrology of China. Acta Geographica Sinica 7:79–85. (In Chinese.)
Tang Y. and Zhang S. 1989. The problems of the control of water quality in economic development of China. Acta Geographica Sinica 44:302–313. (In Chinese.)
Tinachen, Z. 1988. Grassland resources and development of grassland agriculture in temperate China. Rangelands 10:124–127.
Uemura, S., S. Tsuda, and S. Hasegawa. 1990. Effects of fires on the vegetation of Siberian Taiga predominated by *Larix dahurica*. Canadian Journal of Forest Research 20:547–553.
United Nations Center for Science and Technology in Development (UNCSTD). 1990. Report on the International Seminar on Desertification Processes of Contiguous Areas: Science and Technology Policy Issues and Options. Lanzhou, China, 25 June-3 July 1990. (Conference papers in English.)
United Nations Development Program (UNDP). 1992. A Program for the Phase Out of Ozone-Depleting Substances (ODS) in the People's Republic of China. Report of an UNDP mission to China between October 29 and December 22, 1991.
Walker, D. 1986. Late Pleistocene-early Holocene vegetational and climatic changes in Yunnan Province, southwest China. Journal of Biogeography 13:477–486.
Wang M.X., Dai A., Shen R.X., Wu H.B., H. Schultz, H. Rennenberg, and W. Seiler. 1992. $CH_4$ emission from a Chinese rice paddy field. (in preparation).
Wang Tianduo. 1990. An analysis of the problems of water use efficiency in the North China Plain. (Unpublished monograph in English.)
Wohlke, W., Gu H., and Ai N. 1988. Agriculture, soil erosion and fluvial processes in the basin of the Jialing Jang (Sichuan Province/China). Geojournal 17:103–115.
Xie J. and Chen Z. 1990. Water resources in China. Acta Geographica Sinica 45:210–219. (In Chinese with abstract in English.)
Xiwen L. and D. Walker. 1986. The plant geography of Yunnan Province, southwest China. Journal of Biogeography 13:367–397.
Xu W.D. 1982. Correlations between distribution of main forest trees and thermal climate in northeast China. Journal of the Northeast Forestry Institute 4:1–10. (In Chinese.)

Yang, Dongzhen, Ji Xingming, Xu Xiaobin, Fu Jimeng, and Wen Yupu. 1990. The Sandstorm of April 1988: Its synoptic situation and origins. Pp. 197–202 in Global and Regional Environmental Atmospheric Chemistry. Proceedings of the International Conference on Global and Regional Environmental Atmospheric Chemistry, L. Newman, W. Wang, and C.S. Kiang, eds. Springfield, Va: National Technical Information Service.

Yu Zuoyue and Wang Zhuhao. 1990. The method on recovery of forest vegetation in degraded ecosystem in tropical and subtropical waste lowland in Guangdong. *Huanjing Kexue Xuebao* (Journal of Environmental Sciences) 2(3):13–25. (In English.)

Zhang Haifeng, ed. 1984. A Brief Introduction to the Chinese Institutions of Ocean Science and Technology. Beijing: China Ocean Press. 153 pp. (In English.)

Zhang Zhifei. 1991. Strengthening Coordination for More Effective Funding for IGBP. 8 pp. (Conference presentation in English.)

Zhao Ji, Zheng Guangmei, Wang Huadong, and Xu Jialin. 1990. The Natural History of China. London: William Collins & Sons.

Zhao Jianping. 1990. A brief survey of the Chinese Ecological Research Network (CERN). CERN Newsletter 1(1).

Zhu Baoxia. 1991. State pays millions for pollution research. China Daily. 28 August 1991.

Zhu Zhenda, Liu Shu, and Di Xinmin. 1988a. Desertification and Rehabilitation in China. Lanzhou: The International Center for Education and Research on Desertification Control. 222 pp. (In English.)

Zhu Zhenda, Zou Benggong, Di Xinmin, Wang Kangfu, Chen Guangting, and Zhang Jixian. 1988b. Desertification and Rehabilitation. Case Study in Horqin Sand Land. Lanzhou: CAS Institute of Desert Research. 113 pp. (In English.)

Zhu Zhenda and Wang Tao. 1990. Analyzing land desertification trends in China over the past 10-plus years using research on some typical regions. Chinese manuscript received by Acta Geographica Sinica and translated into English in NPRS-TEN-91-013. 5 July 1991, 25 pp.

Zuo Dakang and Zhang Peiyuan. 1990. The Huang–Huai–Hai Plain. Pp. 473–477 in The Earth as Transformed by Human Action, B.L. Turner, II, W.C. Clark, R.W. Kates, J.F. Richards, J.T. Mathews, and W.B. Meyer, eds. New York: Cambridge University Press.

# APPENDIXES

# A

# Overviews of Selected Institutions

## INTRODUCTION

Appendix A contains reports from panel members' visits to selected research institutions during the summer of 1991. They outline the basic organization and resources of the institution, identify researchers and briefly describe research relevant to global change. Visits were chosen from a listing of institutions and individuals involved in global change research that the panel compiled in consultation with the Chinese National Committee for the International Geosphere–Biosphere Program (CNCIGBP) at the outset of this study. Because of time and financial constraints, it was not possible to visit all of the institutions where relevant work is being carried out.

## BEIJING NORMAL UNIVERSITY

### Institute of Low Energy Physics

Beijing Normal University operates a laboratory for elemental analysis by the method of proton-induced X-ray emission (PIXE). The laboratory, with a Van de Graaff nuclear particle accelerator, is managed by Zhu Guanghua and Wang Xinfu, and is administered by the Institute of Low Energy Physics, which is under the leadership of Lu Ting and Wu Yuguang. The laboratory is also actively supported by the university's computer center and its director, Pei Chunli.

## Research Highlights

The laboratory has gained international recognition for its analyses of aerosol particle samples. It has entered into interlaboratory comparison analyses of standard samples with participating laboratories at Kyoto University in Japan and at Element Analysis Corporation in Tallahassee, Florida, with results that demonstrate its ability to produce high-quality analytical results.

Researchers have been assisted through interactions with scientists at other laboratories both within China and internationally. Wang Mingxing of the Chinese Academy of Sciences (CAS) Institute of Atmospheric Physics has advised Zhu Guanghua about aerosol particle sampling and analysis needs in air chemistry research and Wang has made his institute's sampling equipment available to the laboratory. Wang Mingxing brought together Zhu Guanghua and Zhang Xiaoye, a staff scientist at the CAS Xi'an Laboratory of Loess and Quaternary Geology, resulting in collaborative studies of aerosol transport characteristics and publication of their work in international journals.

Zhu Guanghua is collaborating with Yoshikazu Hashimoto of Keio University in Yokohama, Japan and Mitsuru Fujimura and Akira Inayoshi at the Nippon Environmental Pollution Control Center in Tokyo, Japan to measure aerosols through a network of stations in China (interior), Korea, and Japan. Through the Japanese collaboration, computer software has been donated to the university for use in its PIXE laboratory and in teaching programs conducted by the computer center. Additional urban air quality research by the laboratory has been supported by the International Atomic Energy Agency.

Both Zhu Guanghua and Wang Xinfu have attended the triennial international conferences on PIXE and its analytical applications, where they have presented research results. It is likely that the laboratory will become an important contributor to aerosol studies on a global scale.

## CHINA REMOTE SENSING SATELLITE GROUND STATION

The China Remote Sensing Satellite Ground Station, which is administered by CAS, is headed by Wang Xinmin. This ground station is unique in that it is the only Landsat Five, or Thematic Mapper (TM) imagery receiving station in China. The receiving antenna is actually 100 km northeast of Beijing and high-density digital tapes are delivered to the station two or more times a week for processing. Reception is from east of Japan to about 80 percent of China to the west. China needs another receiving station to get imagery for the western 20 percent of the country. This station receives and processes only TM data, and a royalty is paid to EOSAT in the United States. This station does not receive Advanced Very High Resolution Radiometer (AVHRR) data.[1]

TM data are used for mineral exploration purposes, including gold, oil, and coal. China has oil off-shore and in scattered parts of the eastern mainland, and has potentially large deposits in the far western regions (some geologists and others believe that China's oil deposits could rival those discovered in the Middle East). With such possibilities, the Japanese are investing $5.5 \times 10^6$ *yen* in mineral exploration activities in China. Japan is this center's major collaborator.

Major remote sensing application units are the CAS Institute of Remote Sensing Applications, the NEPA Chinese Research Academy of Environmental Sciences, and the Peking University Institute of Remote Sensing Technology and Application (see below). In addition, the station provides remote sensing imagery services to more than 30 other units. For example, the station will be the source of imagery used in a national key remote sensing project in the Eighth 5-Year Plan to monitor and forecast natural disasters and crop production (the Commission for Integrated Survey of Natural Resources [CISNAR] and others will be involved). Such services are not automatic, and many research institutes are not able to afford the cost of remote sensing in their research plans.

About 200 persons staff six technical departments (receiving, processing, photo laboratory, digital processing laboratory, geographic information systems [GIS], and remote sensing applications) and three support departments (planning, logistics, and administration). Skilled personnel include both M.S.- and Ph.D.-level people, many from Tsinghua and Peking Universities. Funding is the specific limitation at the station, especially for keeping up state-of-the-art hardware. Personnel use I-squared software now, but have the capacity to write their own.

Some remote sensing research is going on in the applications department, which is being upgraded in anticipation of the launching of Landsat Six (which is expected to go up soon with upgraded capacity of $15 \times 15$ m pixel in panchromatic). The station is also preparing to handle Synthetic Aperture Radar planned for the Japanese and European remote sensing vehicles. This will change the operations of this unit considerably.

This station has no educational function. Instead, it relies on the remote sensing institute at Peking University (see below) that runs a training center under a coordinated training service for units using remote sensing.

## CHINESE ACADEMY OF METEOROLOGICAL SCIENCES

The Chinese Academy of Meteorological Sciences (CAMS), the research arm of the State Meteorological Administration (SMA), currently has about 500 employees. In addition, CAMS has two graduate schools, one in Beijing and the other in Nanjing. Zhou Xiuji is director of CAMS and Ding Yihui is the deputy director.

According to Ding Yihui, scientists at CAMS working in areas relevant to global change can be found at the Climate Research Center, the Atmospheric Chemistry Center, the Institute of Arid Regions Research, the Tibetan Plateau Institute, and the Beijing Data Center. Currently, about 60 scientists (approximately 20 of which have Ph.Ds) and staff members participate in research subjects related to global change. Annual program funding is about $300,000, excluding salaries.

### Research Highlights

The following are major global change research projects that will be carried out over the coming decade:

- Establish observation networks for detecting global change. CAMS will upgrade existing monitoring networks to monitor changes in climate and large-scale air quality at (a) about 200 standard climate monitoring stations, (b) about 500 agricultural meteorology stations, (c) about 100 acid rain stations, (d) seven standard radiation stations, (e) six total ozone ($O_3$) stations and three $O_3$ sonde stations, (f) six regional air quality stations, and (g) one atmospheric baseline station. Some of these activities are supported by the World Meteorological Organization and they include extensive collaborations with other domestic and international scientists.
- Study the past climate recorded in historical literature. CAMS researchers have studied more than 7,800 pieces of historical Chinese literature dating back to about 1,000 years, and they have established a data set of major droughts and cold and warm periods. In addition, a study of these climate changes and their impact on China's development has been carried out.
- Participate in tropical ocean expeditions and monitoring. CAMS has been active in the Tropical Ocean and Global Atmosphere (TOGA) program since 1984. Major foci are the energy budget of the tropical ocean and atmosphere and the genesis of *El Niño*.
- Model ocean–atmosphere interactions. For ocean–atmosphere coupling dynamic climate models, a three-dimensional global atmosphere model and a six-level oceanic circulation model

have been developed to study ocean–atmosphere interaction problems such as the genesis mechanism of *El Niño*. In addition, a two-dimensional water–energy equilibrium model has been developed to study the mechanism of drought in China.
- Model stratospheric $O_3$ change. A two-dimensional photochemical–radiative–dynamic model has been developed to study stratospheric $O_3$ change and large-scale tropospheric air quality problems.
- Model climate change in East Asia. The U.S. National Center for Atmospheric Research (NCAR) Community Climate Model is used to study climate change in East Asia, including the genesis mechanism of East Asian monsoons and the effects of forest coverage variation on regional climate.
- Study the impacts of climate change on China's development. Population pressures and an extended drought in north China, where a significant amount of China's food grains are grown, are causing harmful changes to agroecosystems and the water supply. Also, destructive meteorological disasters inflict immense impacts on the populace. These problems are being investigated and response strategies are being formulated. In addition, assessment models of impacts of climate change on socioeconomic activities have been developed.

## COMMISSION FOR INTEGRATED SURVEY OF NATURAL RESOURCES

Administered jointly by CAS and the State Planning Commission, the Commission for Integrated Survey of Natural Resources (CISNAR) is responsible for developing and maintaining both past and present data on all the lands and freshwaters of China. With such a national research and data collection mandate, it should be central to any global change research program. CISNAR is headed by Sun Honglie, who is a CAS vice president, member of the CNCIGBP, and chairman of the Chinese Ecological Research Network (CERN). CISNAR has a total of 386 staff, 296 of whom are researchers and technicians, 51 are administrators, and 39 are support staff. The commission also has three ecological field stations (Appendix D).

The view of commission representatives is that CISNAR is central to global change research in China and, rather than having a separate unit devoted to global change considerations, representatives claim to have them built into the work of every project. Such statements belie current limits on understanding of what constitutes global change research or, at the least, indicate a certain expediency in categorizing global change research in its broadest sense.

CISNAR research has an intense ecological flavor. A review of materials provided by CAS about global change research at CISNAR (CAS 1991) reveals interesting work, but, for the most part, research designs do not have any direct connection to global change, and natural resource data collection is not tied to specific research questions. Clearly, the potential exists to develop interesting global change research projects from this work and these data, but it would be overreaching to say it has been done. An historical perspective is said to be included in everything the commission undertakes, and a number of projects include marvelous historical and dendrochronological data.

CISNAR conducts most of the country's large-scale survey research. Much of its work has been expeditionary in character, reminiscent of the former U.S. Biological and Geological Survey. Survey expeditions are multiyear undertakings in large regions such as the Qinghai–Tibet Plateau and Xinjiang Uighur Autonomous Region. CISNAR has completed expeditions in all parts of the country over the last 30 years. Maps at various scales are important products of these expeditions. Because of CISNAR's role in national land use planning, personnel want to develop 15 or 16 case studies to monitor change in different regions.

CISNAR maintains large databases of natural resource information and has plans to increase that capacity. It houses World Data Center (WDC)-D for Renewable Resources and Environment and has a database project with the Ministry of Agriculture (MOA) through their jointly sponsored Integrated Research Center for Natural Resources and Agricultural Development. It has a national climatological database containing information (for the years 1951 to 1980) from about 1,000 monitoring stations. While CISNAR maintains a data management center, it is not clear it has the appropriate hardware and staffing to do the task. They have a VAX computer, a digitizing table, and about two dozen personal computers (PCs). They have no computer work stations.

Extensive plans are being formulated for the establishment of the CERN Synthesis Center (Chapter 4) at CISNAR. The equipment and training planned for CISNAR in CERN plans will address current limitations, particularly in data management and modeling.

The commission plays an explicit educational role by taking on M.S. and Ph.D. students from participating universities and fostering them through research projects. Between 15 and 20 graduate students work at CISNAR.

CISNAR is an extremely important resource for information and understanding of China's land and freshwaters that are important to global change studies. To make its best contribution to such studies,

CISNAR would greatly benefit from increased capacities, including data management, trained personnel, and equipment.

## GUANGZHOU INSTITUTE OF GEOGRAPHY

The Guangzhou Institute of Geography is supported by the Guangzhou provincial government, and is not connected directly to any CNCIGBP activities. In fact, institute staff were not aware of any national plans for global change research. Nevertheless, the institute is doing some work of interest.

The institute is staffed by 110 scientists and is organized into the following departments: physical geography, economic geography, special economic zones, geomorphology and quaternary studies, remote sensing, coastal and river delta resources, laboratory, and administration. A very few undergraduate students from various regional universities conduct research, although the institute does not confer a degree. The institute has eight computers, ranging from IBM XTs to IBM 386s and some Canadian machines. This institute has no GIS capability because of the high cost of hardware and software. The remote sensing department has Eros II hardware and software purchased from a company in Canada. According to institute officials, the department works with TM data, and it can do carbon-14 and chemical analyses.

### Research Highlights

Huang Zhenguo is studying the physical and biological effects of sea-level change on the environment of the Pearl River Delta. Li Pingri, who collaborates with Huang Zhenguo, is doing some very nice historical analyses of coastal changes in the Pearl River Delta. Scientists do not appear to be modeling future changes, although some papers discuss implications; essentially, no predictive work is being undertaken. No collaboration is evident between this coastal zone geomorphology group and Ren Mei-e at Nanjing University, nor with the CAS South China Sea Institute of Oceanology.

Zhong Gongfu is working on a project funded by the National Natural Science Foundation (NSFC) on dike systems, but further information was not available.

## INSTITUTE OF ATMOSPHERIC PHYSICS

The CAS Institute of Atmospheric Physics is a leading center for research on a broad range of atmospheric physics, including regional and global atmospheric circulation, boundary layer physics, pollution meteorology, tropospheric physics, and atmospheric chemistry. Zeng Qingcun, director, is well known internationally for "Zeng's model" for general circulation. Ye Duzheng, director *emeritus*, is internationally known for his work on atmospheric circulation in eastern Asia, particularly for his work on the role of the Qinghai–Tibet Plateau in general global circulation. The institute has about 550 staff members, 14 laboratories, a computing center, a postgraduate division, a postdoctoral program, and an atmospheric observatory at Xianghe.

Since 1986, scientists have been carrying out research in nine areas: climate change and prediction, medium-range weather forecasting, mesoscale dynamics and nowcasting, atmospheric environment, acid rain, middle atmosphere, geophysical fluid dynamics, atmospheric physics and chemistry, and global change. In general, two research areas are of particular relevance to the global change studies—climate modeling and climate diagnostics. The institute is a primary collaborating institute under the U.S. Department of Energy (DOE)-Chinese Academy of Sciences Joint Research on the Greenhouse Effect (Appendix C).

## Research Highlights

A hierarchy of climate models for the atmosphere and oceans has been used in climate simulations, and more models are under development. The institute's 2-level atmospheric general circulation model (GCM) has been developed since 1987 and is used extensively for climate studies, such as the Asian monsoon, seasonal abrupt changes in atmospheric circulation, severe cold summers, low-frequency oscillations, and teleconnection. The model is also used to participate in the international GCM intercomparison program organized by DOE for understanding differences among GCMs (Appendix C). A multilayer ocean GCM and a multilayer atmospheric GCM are separately under development, with the ultimate goal of studying the ocean–atmosphere interaction. As part of model validation study, these models will also be used to examine the paleoclimate.

Climate diagnostics mainly focus on seasonal, interannual, and long-term climate variability, in particular, monsoon variability. The main purpose of this focus is to understand the physical mechanisms of weather process, such as the formation, propagation, and anomaly of planetary waves, teleconnection, low-frequency oscillation, air–sea coupled oscillations and the effects of *El Niño*–Southern Oscillation (ENSO) on the general circulation and the climate of China. In recent years, special emphasis has been placed on the 14- and 40-day oscillations, the relationship between the Asian monsoon and summer rainfall in the Yangtze River Valley and the relationship between the Chinese and Indian monsoons. The objective is to improve understanding of the climate system for long-range forecasting and short-term prediction.

The National Key Laboratory of Numerical Modeling for Atmospheric Sciences and Geophysical Fluid Dynamics (LASG), set up in 1985, hosts visiting scientists from all over the world. LASG supports climate, ecology, and environmental research by using data sets and computer-assisted analyses. Main topics are the development of four-dimensional dynamic models, data sets, and methodologies and protocols for handling large volumes of data.

The LASG data center holds more than 200 gigabytes, and the volume of data is growing at more than 1 gigabyte per year. The LASG has a CONVEX mini-supercomputer, Silicon Graphics work station, and numerous micro-computers. Data holdings include basic climate variables, trace gas, land-surface properties, ocean variables, past climate change variables, and global and national carbon dioxide ($CO_2$) emissions.

The institute publishes *Scientia Atmospherica Sinica* (*Daqi Kexue*, quarterly, in Chinese), *Advances in Atmospheric Sciences* (quarterly, in English), and the *Annual Report of the Institute of Atmospheric Physics*.

## INSTITUTE OF BOTANY

The CAS Institute of Botany is located in pleasant surroundings on the grounds of the Beijing Zoo. The institute is under the leadership of Zhang Xinshi (also known as Chang Hsin-shih and David Chang), who replaced the distinguished Tang Peisung upon his retirement. The institute is large, effectively managed, and covers a wide array of botanical sciences. It has sections dealing with phytochemistry, cytology, nitrogen fixation, photosynthesis, plant physiology, and plant ecology/geobotany. It has the largest herbarium in Asia, with collections predating the 1920s. The professional staff is 580, with a support staff of 200. Approximately 50 M.S. and 20 Ph.D. students are working in the institute. The institute administers three CERN sites: (1) the Inner Mongolia Grassland Ecosystem Experiment Station, (2) Beijing Forest Ecosystem Experiment Station, and (3) Maousu Ecology Experiment Station. The institute is working with the CAS Lanzhou Institute of Plateau Atmospheric Physics and the CAS Lanzhou Institute of Glaciology and Geocryology to establish a research station on the Qinghai–Tibet Plateau.

## Research Highlights

The institute is involved in interdisciplinary research programs within China and has active international programs with laboratories in Australia, the United Kingdom, and the United States. A major global change study is centered in the Laboratory for Quantitative Ecology, which is described in detail in other sections of this report that detail activities in the Global Change and Terrestrial Ecosystems (GCTE) Core Project, land cover change, and biogeochemistry (Chapters 4 and 5). Zhang Xinshi will be working with Zheng Du, CAS Institute of Geography, on a multi-institute, multidisciplinary project, "The study of the Origin, Evolution, Environmental Change and Ecosystems of the Qinghai–Tibet Plateau," which is funded by the State Science and Technology Commission (SSTC).

Other sections of the institute offer important academic support for the global change effort. For example, the photosynthesis research group, well known for its work on chloroplast membrane structure and biophysics, is planning to participate in climate–vegetation studies by taking part in the construction of models of photosynthesis, regulation of stomatal conductance, and respiration.

Institute researchers are clearly taking an important role in the Chinese global change program and are anticipating an expanding role, particularly in the study of climate–vegetation interactions.

## INSTITUTE OF GEOGRAPHY

The Institute of Geography is large and has an extensive mission, including studies of climate, natural resources, economics, and paleoclimate. Under joint sponsorship of CAS and the State Planning Commission (SPC) since 1986, the institute serves as a principal source of geographic data on physical, biological, and economic resources of China, as well as maintaining basic research on natural and cultural resources, information science, and global change.

The institute, under the leadership of Zuo Dakang, has 13 departments, four laboratories, and the National Key Laboratory of Resources and Environment Information Systems (LREIS), which has links to other domestic and foreign institutes. The institute administers two CERN stations: the Yucheng Integrated Experiment Station in Shandong Province focuses on water balance involving moisture exchange in the soil–plant–atmospheric system, and the Beijing Agroecology Experiment Station focuses on the regularity of energy transformation and substance migration of farmland and agroforestry systems in North China and the interrelation of growth and yield formation of crops and woods.

LREIS is equipped with world-class GIS facilities, image processing, cartography, and photointerpretation. The laboratory uses ARC/INFO, a GIS software developed in the United States, and serves as the Asian distributor for this software package. The GIS system is operated on recent Digital Equipment Corporation work stations, which are in the process of being upgraded. LREIS has the ability to produce high-quality maps from aerial photography, Landsat, or other satellite data sources. The laboratory's hardware and software are comparable to those found in U.S. institutions.

LREIS personnel are very much part of the international GIS community, and LREIS has also hosted several recent conferences about GIS. Institute personnel are well trained and experienced in a range of disciplines. Several distinguished scholars are on staff, as well as strong technical personnel. The group is well connected internationally and has considerable exchanges with American and other institutions.

Despite the applied roots of the institute's data activities, it could be a larger player in IGBP's Data and Information Systems (DIS) activities in China and this type of collaboration should be encouraged. The level of China's current participation in DIS is similar to organizations in the United States, for example, the U.S. Geological Survey, and other nations that are also just beginning to participate. The limited role of LREIS, a GIS national laboratory, in Chinese global change research, is due, in part, to its mandate to support planning and development activities and to competition for its overcommitted resources. The institute does not appear to be involved in the WDC-D effort, although it has recently established a specific department to support the CERN subcenter for hydrology (Chapter 4).

## Research Highlights

Studies related to global change are conducted mainly in the Department of Climatology and Department of Hydrology. The Department of Climatology studies basic laws of climate formation and evolution and analyzes interactions of various physical processes and changes in past climate periods, maritime climate, monsoon climate, urban climate, near-ground physical climate, and phenology. Recent studies on climate change have concentrated on the East China Plain and the Qinghai–Tibet Plateau.

China possesses a vast resource of historical writings that have frequently been studied by historians, philosophers, and social science scholars. These writings contain climate information that was recorded in different types of historical sources: local gazettes, official dynastic histories, personal diaries, ancient literature, and memos to the emperors. It is possible to extract and establish climate series of precipitation, harvests, disasters, and phenology by studying and analyzing these writings. These data also permit historical analysis of the impact of climate change on regional resources, such as water supply, as well as social and economic activities.

Chu Kezen was the first one to use historical writings to establish the history of temperature fluctuation during the last 5,000 years in China. In recent years, under a DOE-CAS joint project (Appendix C), Zhang Peiyuan has been leading a group that is extracting key information on China's past environment from several sources, including the Imperial Archives and tree rings. Most of these historical data have been stored on computer. Some time series of past climate change have been reconstructed: dryness and wetness variation in Beijing (1260–1979), dryness and wetness variation in Luoyang (1000–1979), and temperature fluctuations in Nanjing, Suzhou, and Hangzhou (1724–1980).

Tree-ring sampling, analysis, cross dating, and standardization are carried out in the tree-ring laboratory, thus providing climate information and calibration with historical and instrumental data.

The Department of Hydrology conducts research on slope runoff on the Loess Plateau and on the transformation of *Sishui* (surface, ground, soil, and atmospheric water) on the Huang–Huai–Hai Plain. Hydrographies of rain type, drainage morphology, and ground conditions are simulated in laboratories. This latter work, which is referred to as the "soil, plant, atmosphere continuum," is labeled as a major new trend in the institute's hydrological research. In recent years, researchers have also studied the water transfer from the Yangtze River to the Huanghe River basin and water resources along the coastal regions.

The institute will be involved in the multidisciplinary project, "The Origin, Evolution, Environmental Change and Ecosystems of the Qinghai–Tibet Plateau," which will be conducted from 1992–1996. Zheng Du, a physical geographer, is a lead scientist for the ecosystem portion of the project along with Zhang Xinshi, CAS Institute of Botany. Zheng is a member of the expert committee that is in charge of implementing the project.

Institute activities represent an exciting opportunity to link natural science and human impact studies. The scholarly activities of the institute in paleoenvironments are excellent and innovative, coupling documentary and geophysical evidence in a unique way. Global change research enjoys the strong interest and involvement of the institute's senior scientists.

## LANZHOU INSTITUTE OF GLACIOLOGY AND GEOCRYOLOGY

The mission of the CAS Lanzhou Institute of Glaciology and Geocryology is to study the glaciers, ice sheets, and frozen soils of China. Research activities include basic research on ice and snow hydrology and physics and engineering of frozen soils, for example, road building on permafrost and development of monitoring instrumentation for use in harsh environments, including cold, high-altitude, and snow-covered areas.

Institute staff, under the leadership of Chen Guodong, number approximately 400, of which more than 300 are scientists and technicians. The institute contains several scientists who have overseas training in well-regarded U.S. and European laboratories and who have chosen to return because of the unparalleled access the institute provides to mid-latitude glaciers. This high rate of returning students is quite unusual and noteworthy.

Research facilities include the National Key Laboratory for Frozen Soil Engineering, the Tianshan Glacier Research Laboratory at 4,000 m and the Tianshan Glaciology Research Station at 5,000 m (located in the Tianshan Mountains near Urumqi in Xinjiang Uighur Autonomous Region), and, jointly with the CAS Lanzhou Institute of Plateau Atmospheric Physics, the new logistical base station and comprehensive observatory located respectively in Golmud and Wudaoliang in Qinghai Province. The institute houses and maintains WDC-D for glaciology and geocryology. Plans are under way for the construction of an ice core laboratory.

Laboratories are well equipped, including instrumentation for elemental analysis by ion chromatography, liquid water content of frozen soils, and other analyses that are standard in these areas of research. The institute includes unique facilities for carrying out measurements on frozen samples, especially mechanical analyses. Funding has been secured for a mass spectrometer to allow analysis of the isotopic composition of glacier air and for studies of paleoclimate. When this instrument is delivered, the institute will have world-class facilities for the study of paleoclimate and atmospheric chemistry, a key subject in global change research. The laboratory also has modern computing facilities, including a SUN 4 work station, which was obtained via a collaborative project with the University of California, Santa Barbara.

Historically, the scientists from the institute have pursued international collaboration and foreign funding aggressively. Nearly 50 percent of research funding is soft money, with half of that grant funding coming from abroad. The institute's chief global change activity is in hosting a series of international expeditions exploring the chemistry of ice cores from mid-latitude glaciers, an avenue of research first explored with U.S. partners, and now carried out with colleagues from many countries, including Japan, the former Soviet Union, and several European nations. The institute organizes in-country logistics, which can be quite challenging given the remote locations and difficult terrain in which the glaciers and ice sheets of China are found.

## Research Highlights

Global change research is classified into four areas: (1) ice core studies, (2) snow cover studies, (3) glacial mass balance snowline, glacial discharge, and climate change, and (4) Antarctica.

### *Ice Core Studies*

Ice core studies began in 1982 through a cooperative program with Australian researchers, and currently, active cooperative projects with Americans and Japanese are ongoing. From 1984 through 1987, Lonnie Thompson, from Byrd Polar Research Center at Ohio State University, Xie Zichu, and Yao Tandong conducted studies of ice cores from the Dunde ice cap in the Qilianshan Mountains that form the border between Gansu and Qinghai Provinces. Results of this work show more pronounced temperature fluctuation in the past 50 years than in the previous 10,000.

As part of a paleoclimate study of global climate variability, institute researchers and Lonnie Thompson currently are studying the Guliya ice cap in the western Kunlunshan Mountains in southern Xinjiang Uighur Autonomous Region. Results show that this ice cap has the thickest ice (350 m) and lowest temperatures outside of polar regions, and that it is the highest (6,700 m above sea level) and the largest (320 $m^2$) subtropical ice cap in the world. Ice core samples will be analyzed for past climate, atmospheric gas, and dust composition. Samples will also be taken from low-latitude, high-elevation sites in Peru. It is expected that the Guliya ice cap will contain the best record of the relationship of low-latitude, high-elevation climate to the record of monsoon variability in southern Asia, which has bearing on the relationship between ENSO and monsoon activity (NSF 1991).

Researchers are involved in a major multidisciplinary project, "The Origin, Evolution, Environmental change, and Ecosystems of the Qinghai–Tibet Plateau." Shi Yafeng (who also has a research appointment at the CAS Nanjing Institute of Geography and Limnology) is working with Li Jijun of the Department of Geography at Lanzhou University and Li Bingyuan at the CAS Institute of Geography to study environmental changes during the Late Cenozoic Era. Subcomponents include the ice core study mentioned above, lacustrine core drilling (in cooperation with researchers from the CAS Nanjing Institute of Geography and Limnology [see below] and the CAS Guiyang Institute of Geochemistry), and six natural profiles will be taken from along the edges of the plateau.

### Snow Cover Studies

Snow cover studies concentrate on understanding the influence of climate and environmental change on snow cover and feedbacks to climate and the environment. One of the major cooperative projects between Li Peiji and George Kukla of Columbia University has been to investigate the influence of climate change on snow cover in western China, including the study of the relationship between the greenhouse effect, volcanic eruptions, ENSO, and the Indian monsoon on snow cover. Xie Zichu and M. Kotlyakov from the former Soviet Union have been conducting a similar project. Yang Daqing and Jeff Dozier, University of California, Santa Barbara, are studying the spatial and temporal distribution, surface energy transformation, and melting process of snow cover by using field observations at the Tianshan Mountain research stations and remote sensing data.

### Glacial Mass Balance Snowline Studies

Glacial mass balance snowline, glacial discharge, and climate change is a research area that has been pursued since the founding of the institute. Currently, Liu Chaohai is the lead scientist for these studies. Several projects are under way: features and fluctuations of glaciers and snow cover in the Tianshan Mountains (a cooperative project between China and the former USSR), the influence of climate change on water resources in northwest China and future climate trends, the relationship between cryosphere, hydrosphere, and atmosphere on the Qinghai–Tibet Plateau (a cooperative project between China and the former USSR), and measuring existing glacier fluctuation under phase four of the UNESCO International Hydrology Project.

### Antarctic Studies

Glaciology studies in Antarctica were initiated through cooperative projects with Australia. One of the primary investigators, Qin

Dahe, gained international recognition in 1990 as a member of the Trans-Antarctic Expedition. Samples taken during the trek are being analyzed. Around China's Great Wall Antarctic station an ice core study of Holocene Epoch environmental evolution is being conducted in cooperation with Uruguay and the United States.

The institute occupies an important role in global change research in China because it controls access to an important research resource—mid-latitude glaciers—which has brought it funding and extensive international collaboration. Excitement stemming from this role has encouraged the development of a young and highly motivated staff. While some potential U.S. collaborators have encountered problems reaching agreement about cost sharing, overall cooperation with the United States remains strong, due in part to the high mutual regard of individual scientists.

## LANZHOU INSTITUTE OF PLATEAU ATMOSPHERIC PHYSICS

The CAS Lanzhou Institute of Plateau Atmospheric Physics is responsible for the investigation of the atmospheric sciences in the Qinghai–Tibet Plateau region, and in areas where the climate is influenced by the plateau. The institute conducts a broad range of studies in atmospheric modeling, numerical weather prediction, climatology, boundary layer meteorology, clouds and precipitation, atmospheric electricity, and radar meteorology. Researchers are making a strong effort in bioclimatology, and this is the principal institute for the Sino–Japanese Atmosphere–Land Surface processes cooperative (HEIFE) experiment (Chapter 4), a bilateral interdisciplinary investigation of the coupling of land surface hydrology, boundary layer dynamics, and mesoscale meteorology.

Under the open and energetic leadership of Guo Changming, the institute is a fairly compact institute of 290 total staff, of whom 60 percent are scientists. It has a strong cohort of young and competent scientists (most staff are under 35), and a graduate program with 20 students. However, it is difficult to retain the brightest young scientists because institutes from less remote areas frequently are able to recruit successfully from this institute. It has a technical support section that produces and maintains instrumentation, much of which is quite impressive.

The institute is housed in functional quarters in Lanzhou. Computing at the institute is limited to PCs and a shared VAX 11/780 computer. Simulation modeling and numerical weather prediction is severely constrained by the lack of computing facilities. Most software used in modeling, model diagnosis, and data analysis are written in-house. The institute has field facilities for atmospheric electricity studies and several field stations, including one it sponsors jointly with the CAS Lanzhou Institute of Glaciology and Geocryology near Golmud on the Qinghai–Tibet Plateau that supports measurement stations at altitudes from 2,800 to 5,000 m. Facilities for micrometeorology and boundary layer meteorology are excellent, including an acoustic sounder, profiling towers, an eddy correlation system, tethersondes, and radiation instruments. At the HEIFE experiment site, instrumentation support personnel have developed and deployed a shortwave radio telemetry system for relaying remote micrometeorological observations to a central receiver and computer.

## Research Highlights

While core activities are in the areas of traditional meteorological and climatological research, the institute supports a number of activities that are of direct or supporting importance to global change science. The HEIFE experiment, which addresses land–atmosphere coupling, is of central importance to the Chinese global change program. Supporting activities include ones in synoptic scale and dynamic meteorology, especially in relation to the influence of the Qinghai–Tibet Plateau.

In contrast to general panel observations, planning for the HEIFE experiment involved some coordination of modeling and experimental design by using a mesoscale model to determine whether surface fluxes at experimental sites were representative of the region as a whole. Efforts to integrate modeling into experimental design as part of the HEIFE experiment are very innovative and the potential for leadership in climate system modeling from this institute seems high, despite its remote location and limited computing facilities.

Panel members were surprised to find out that this institute was not included in the CERN network. It appeared that it could contribute to the network in several ways, as well as be able to make good use of investments planned for CERN stations. The panel understood that one of the problems with participation was the institute's lack of a permanent experiment site. However, the Linze site for the HEIFE experiment has and will continue to offer valuable data and be a site of scientific interest. Furthermore, this institute's work fills an important niche in the distribution of CERN sites around the country. The institute's overall strengths make it a sound potential partner in developing a better integration of biophysics and meteorology with ecology.

## NANJING INSTITUTE OF ENVIRONMENTAL SCIENCE

The Nanjing Institute of Environmental Science occupies attractive quarters with modern facilities on the eastern side of the city. Under the direction of Zhou Zejiang, this is an institute under the National Environmental Protection Agency (NEPA). This particular institute specializes in rural environmental problems and it appears to be staffed by a group of sharp young scientists.

### Research Highlights

Zhou Zejiang listed six activities considered germane to global change research: (1) preserving biological diversity; (2) deforestation and reforestation; (3) desertification; (4) ecologically sound agricultural practices; (5) biogenic gas flux related to agricultural practices; and (6) biogas production.

A methane ($CH_4$) flux study is ongoing at a research area about 50 km from Nanjing, the results of which were presented at a NEPA–CAS–U.S. Environmental Protection Agency (EPA) symposium in Beijing in May 1991 (Li et al. 1991). The basic question being addressed is the effect of rice culture methods on $CH_4$ flux—with a goal being to reduce that flux. Results show large differences, but insomuch as the fertilization and straw treatments were confounded by irrigation treatments in an unbalanced factorial design, the factors underlying the differences cannot be isolated.

Research groups at the CAS Nanjing Institute of Soil Science (Yang Linzhang and Dong Yuanhua, PIs) and this institute (Li et al. 1990) are on the verge of measuring nitrous oxide ($N_2O$) flux simultaneously but are blocked, in part, by a lack of reliable calibration gases. This would be easy to overcome through U.S. collaboration.

## NANJING INSTITUTE OF GEOGRAPHY AND LIMNOLOGY

The CAS Nanjing Institute of Geography and Limnology is under the leadership of Tu Qingying. It is housed in a lovely new building located on a campus with other CAS institutes, including the Nanjing Institute of Soil Science. This institute supports about 20 M.S. students, two Ph.D. students, and a couple of postdoctoral fellows. This may be the largest limnology institution in China, and researchers work on lakes all over the country. This limnological work has a strong component of historical analysis that is germane to the IGBP Past Global Changes (PAGES) Core Project. These lakes could also serve as sensitive monitors of change as well.

APPENDIX A

The institute holds one of China's repositories of ancient written records. This treasure is being analyzed by Chen Jiaqi for flood records of major Chinese rivers over the last 1,500 years (Chen 1986, 1987, 1989, 1991, Undated). Another historical approach is under way by Qian Junlong to analyze tree rings chemically for trace metals (lead, manganese, copper, and cadmium) over time and with respect to soils (Qian Undated, a,b,c). It is of additional interest that she is using an inductively coupled plasma emission spectrometer for this work.

A study of the tendency and impact of climate and sea-level change in China involving 20 institutes is led by Shi Yafeng (Shi 1991). Scientists are reconstructing climate changes over the past 10,000 years and predicting future changes by modeling climate change under various conditions, for example, elevated $CO_2$ and trace gases.

Wang Suming, who works in the Lake Sediment and Environment Laboratory will be working with Wen Qizhong of the CAS Guiyang Institute of Geochemistry to study lacustrine core drilling as part of the component to study environmental change in the Late Cenozoic Era in the SSTC-funded project, "The Origin, Evolution, Environmental Change and Ecosystems on the Qinghai–Tibet Plateau," which is starting in 1992. Researchers from the Limnological Research Center at the University of Minnesota are also involved in this component.

Other projects were presented as part of Chinese global change research: (1) a riverine hydrology simulation model for two major rivers that takes into account changes in catchment characteristics and ground water and leads to an assessment of the relationships with sea-level rise and effects on crops (Pan Liangbao, PI); (2) coastal zone geomorphology for China's central eastern coast) (Shi Shaohua and Zheng Changsu, PIs); (3) historical and contemporary hydrometeorology of Chinese lakes, particularly in the northwest and Qinghai–Tibet Plateau. (Zhang Xuebin, PI); (4) climate cyclicity from various records but especially lake sediments (Pu Peimin, PI) (Pu 1983, Undated); and (5) remote sensing and GIS.

The institute has some nice equipment and good facilities. In the GIS laboratory, an official stated that the institute cannot afford to buy commercial GIS software, and students have written their own.

Global change research is still in the planning stage. Furthermore, funding is limited and this lack of money has prevented the implementation of projects. Officials would welcome international collaboration in order to be able to start global change research.

Overall, this institute offers particular strength in limnology and some kinds of historical analyses of environmental changes appropriate to PAGES. It may offer capabilities in GIS and remote sensing, coastal zone geomorphology and sedimentology relevant to the proposed Land–Ocean Interactions in the Coastal Zone (LOICZ) Core Project, hydrologic modeling, and climatology. However, further communication and confirmation is required. The institute may offer some capacity in the estimation of energy and mass fluxes between land and atmosphere, but again, this awaits further confirmation.

## NANJING INSTITUTE OF SOIL SCIENCE

The CAS Nanjing Institute of Soil Science, directed by Zhao Qiguo, is the largest institution in China devoted to soil science, and may be considered the leading center for soil science in the country. Basic research focuses on studies of soil genesis and classification, characteristics of soil distribution, the physical, chemical, and biological processes of soils, and the relationships between soil environmental conditions and plant growth. The institute has almost 500 personnel, almost 400 of whom are scientists, engineers, or technicians, and approximately 20 M.S. and five to ten Ph.D. students.

The institute has ten departments, three experimental stations, and houses the National Key Laboratory of Material Cycling in the Pedosphere, which is open to visiting Chinese and foreign scholars. The buildings are attractive and well maintained inside. The laboratories are clean, orderly, and equipment includes an X-ray spectrograph housed in an air-conditioned laboratory, a minimal Swedish autoanalyzer, and a Perkin-Elmer atomic absorption spectrophotometer.

The following are three field stations the institute administers: (1) Fengqiu Comprehensive Agroecology Experiment Station in Hunan Province on the Huang–Huai–Hai Plain, where wheat, cotton, corn, and other crops are grown; Yingtan Red Soil (ultisols) Hill Experiment Station in Jiangxi Province, where rice, winter wheat, peanuts, corn, and sorghum are grown; and (3) Taihu Agroecology Experiment Station, where rice, vegetables, and fish are grown.

Much research has focused on developing methods to improve soils by using fertilizers and salt-reducing techniques, soil genesis and classification, soil properties, soil information systems, and other methods for improving agricultural productivity of the land. Additionally, the institute is responsible for conducting a national soil survey.

### Research Highlights

The institute is involved in three areas of global change study: (1) the study, prediction, and control of soil changes associated with

APPENDIX A

land use and climate change; (2) the execution of germane experiments and monitoring of change at its stations; and (3) biogenic trace gas research.

Yang Linzhang and Dong Yuanhua are measuring $CH_4$ and $CO_2$ from rice paddies at the Taihu station. This area used to be under a double cropping system in which two rice crops per year were grown. Due to a shortage of labor in the countryside, this practice has been discontinued. Instead rice is grown in the summer months (May to September) and wheat or rice from late October to May. This change in cropping systems is important to researchers estimating $CH_4$ flux from rice paddies. The site has an automatic sampling system and they are measuring $N_2O$. This project will continue through 1995.

Three levels of variability are present in this $CH_4$-producing system: (1) diurnal, (2) events within the rice growing period, and (3) the annual cycle of the cropping system. Another aspect of this system is the fate of the straw, which is either returned to the field, burned in the field, burned domestically as fuel, or fed to animals. It is important to add this last variable when assessing the entire production system.

According to CAS (1991), under a United Nations Environment Program (UNEP) project to produce a global map and assessment of soil degradation, institute researchers produced maps for China and North Korea. Building on this experience, scientists are studying the effect of global climate warming on soils. Expected results include a 1:10,000,000 map of soil change, a soil–terrain database, the development of models for global soil change, and a book on the impact of soil change.

Researchers need a spatial database for extrapolating results for this province, not to mention all of the paddy areas of China. GIS support is available at the institute through Lin Guangsong and Yang Xiangheng and at the CAS Nanjing Institute of Geography and Limnology located just yards away.

## NANJING UNIVERSITY

### Department of Geo and Ocean Sciences

The Department of Geo and Ocean Sciences at Nanjing University is home to an ambitious ongoing research program involving sediment discharge and dynamics, coastal and continental shelf geomorphology, and changing levels of the land relative to the sea for the three major river systems of China (Yangtze, Huanghe, and Pearl). This program is led by the renowned scientist, Ren Mei-e (Ren 1991a), who has numerous collaborators and students involved in this project at the university and at other institutions.

This project fits in exactly with some LOICZ objectives. Ren has the beginnings of an expanded proposal for an integrated study of land cover change, river hydrology and geomorphology, and human and economic impacts in the coastal zone of the Huanghe River (Ren 1991b). This proposal could be a model for global change research in China. He is approaching these phenomena through historical analysis now but was receptive to making it predictive.

## NORTHWEST INSTITUTE OF SOIL AND WATER CONSERVATION

The CAS Northwest Institute of Soil and Water Conservation is located about 90 km from Xi'an in Yangling in Shaanxi Province. It is a large institute, with a total staff of 372, including 296 scientific and technical personnel, 81 of whom are scientists with research qualifications. Li Yushan is director of the institute. Because of a last minute mix-up, panel members were not able to visit the institute as planned, but two members of its staff, Tang Keli, director of the institute's soil erosion laboratory, and her associate Wang Binke, met with panel members in Xi'an and provided printed materials and verbal descriptions of their programs.

The institute is dedicated to studying soil and water conservation techniques on the Loess Plateau. In anticipation of climate change on the Loess Plateau, research focuses on changes in agricultural practices that will be necessary.

The institute also conducts studies of past climate change in this region of China. The institute is interested in cooperative research with other scientists, especially in the United States, on studies related to IGBP. In fact, Tang Keli was involved in a bilateral meeting of scientists in 1988, organized by the CSCPRC, to initiate cooperative projects on global change. The development of collaborative projects has been limited to date. Although progress has been slow, researchers at this institute remain enthusiastic about developing international cooperation.

Possible topics for collaboration include soil erosion processes, economic benefits of soil conservation, changes in vegetation in response to soil erosion, geologic record of soil erosion, effects of human activity on soil erosion, and developing a GIS database on Huanghe River sediments.

The institute operates seven field experiment sites in the Loess Plateau region. Lists of equipment contain some 2,000 instruments, some of which are large and include a multichannel analyzer, an inductively coupled plasma quantorecorder, rainfall simulators, remote sensing image processing system, and soil physical–chemical analyzers. The library holds over 85,000 volumes, including 1,100 periodicals, 2,400 journals, and 4,500 publications.

Publication of scientific papers by the institute is found principally in Chinese journals, some of which are in English. Publication in international journals also occurs. The reputation of the institute for the quality of its published research is generally good.

## PEKING UNIVERSITY

### Research Highlights

A number of relevant research projects have been or are ongoing at Peking University in areas related to global change. Like many other Chinese institutions, work on paleoclimate outweighs other types of research. Paleoclimate studies include studies of loess in the Miaodao islands of Shandong Province in the Pleistocene by Cao Jiaxin and of loess in Shanxi, Shaanxi, and Ningxia Provinces in the Quaternary Period by Wang Nailiang. Two studies of glaciation in the Pleistocene Epoch are ongoing by Liu Gengnian and Cui Zhijiu. A project looking at past human interactions with the environment on the Beijing Plain in the past 5,000 years is ongoing under Hou Renzhi and Xu Haipeng. Han Mukang has completed work on a study of the interaction between environment and sea level.

Historical studies of climate change have included the use of travel diaries by Yu Xixian, tree rings by Liu Jihan, precipitation records from the past 500 years by Wang Shaowu, and a reconstruction of the unusual cooling in the Dali area of Yunnan Province during the last half of the thirteenth century by Yu Xixian.

Studies of climate change by using modern measurement techniques have been completed. Wang Shaowu studied the long-term cycles in air temperature and their relation to global atmospheric circulation. With Zhao Zhongci, he has studied long-term precipitation cycles. Zhao has used numerical modeling of $CO_2$ to study impacts on global climate change. Tang Xiaoyan and Shao Kesheng have studied the effect of $CH_4$ and acid rain on atmospheric chemistry.

As seen below, the university has a remote sensing center that has been used to study the impacts of climate change. For example, Cui Haiting has completed a study of changes in vegetation cover in ecological transition zones in North China and Fan Xinqin has studied urban heat islands in Beijing. Other climate impact studies include studies of water resources by Cui Haiting, desert landscape change by Cui Haiting and Huang Wenhua, and changes to the seacoast around Qinhuangdao from the Holocene Epoch to the present by Zu Haipeng.

## Center for Environmental Sciences

The Center of Environmental Sciences, under the leadership of Tang Xiaoyan, professor of environmental chemistry, is the major organizational unit under which scientists specializing in chemistry, physics, and mathematics join together in projects in environmental sciences in cooperation with SSTC, NEPA, and the Beijing government. The Department of Technical Physics and the Department of Geosciences also have some faculty working in environmental sciences.

Research fields focus especially on atmospheric photochemistry and acid precipitation. Facilities include a photochemical smog chamber, ion chromatography, and other instruments. A great advantage of conducting research at Peking University is the wide range of opportunities to cooperate with different academic departments. Because of the high prestige of the university, many excellent graduate students carry out thesis investigations, adding to the versatility of research activities.

Under a cooperative exchange arrangement with the University of North Carolina, students may study there for up to 1 year and American scientists may spend up to 3 months in China lecturing, teaching, and conducting research. Results of this program, which is sponsored by EPA (William Wilson, PI), has resulted in advanced training in analytical methods for air pollution research and exchanges of software and modeling procedures.

Under a tripartite agreement among the center, the University of Michigan, and the Russian Academy of Sciences, annual workshops will be arranged to promote international interdisciplinary research and training on global change. No workshops had been held as of the summer of 1992.

Though it sponsors no major journal of environmental science, Peking University is home to influential reviewers of papers in important national journals.

## Institute of Remote Sensing Technology and Application

Depending on which sponsoring organization is identifying it, this unit can be known as the Institute of Remote Sensing Technology and Application (Peking University), National Remote Sensing Center Training Department (SSTC), or the Chinese Universities Remote Sensing Center (State Education Commission [SEDC]). The institute was established in 1983, and has a staff of about 30 to 35 professors, two associate professors, and four senior engineers. Four codirectors oversee different aspects of the institute's work: Cheng Jicheng (research), Chen Kai (administration), Ma Ainai (education), and Xu Xiru (laboratory). A GIS and a geographic expert system are run on PCs. A satellite ground station receives AVHRR data from National Oceanographic and Atmospheric Administration (NOAA) satellites. Research by using remote sensing techniques is usually geographic in nature and is mainly concerned with water and soil resources. Researchers also carry out geological surveys of ore deposits.

The institute is a training center for institutions using remote sensing imagery, which has reduced the need for individual administrative systems to build up their own training capacities. Furthermore, the center has been the site for regional training programs, for example, a United Nations program for GIS training for the Asian Pacific region.

## QINGDAO INSTITUTE OF OCEANOLOGY

When established in 1950, the CAS Qingdao Institute of Oceanology was the first of its kind in China. Its unique contribution, especially its work on marine algae, marine sedimentation, shallow water circulation, experimental marine biology, cultivation technology, and marine resource development, has laid the foundation for the development of Chinese marine science studies. Directed by Qin Yunshan and with a staff of 1,079, it is the largest multidisciplinary oceanographic institute in China. In addition, Qingdao is home to other major oceanographic institutions such as the State Oceanographic Administration's (SOA) First Institute of Oceanography and the Qingdao Ocean University.

The institute has three open laboratories (that accept visiting Chinese and foreign scholars) for studies of marine biology, ocean circulation and air–sea interaction, and ecological toxicology studies. It has nine major departments, namely, the Department of Physical Oceanography, Department of Marine Geology and Geophysics, Department of Marine Chemistry, Department of Marine Environmental Science, Department of Marine Botany, Department of Marine Invertebrate Zoology, Department of Marine Vertebrate Zoology, Department of Marine Experimental Zoology, and Department of Marine Technology and Instrumentation.

The institute also has supporting facilities, including a central laboratory, Department of Scientific and Technological Information, three research stations—Yantai Station, Xiamen Station, and Huangdao Mariculture Experiment Station (also a CERN station; Chapter 4), ocean-going research vessels, and an administrative office.

## Research Highlights

Hu Dunxin, who heads the CNCIGBP's working group on ocean flux studies, is involved in Chinese Joint Global Ocean Flux Study (JGOFS) activities (Chapter 4). In general, research is centered on resource development and management through comprehensive surveys and studies of China's marine environment and resources. Given what is known about this institute, follow-up is encouraged to determine more extensively its role in oceanographic global change research.

## RESEARCH CENTER FOR ECO-ENVIRONMENTAL SCIENCES

The CAS Research Center for Eco-Environmental Sciences (RCEES) was established in October 1986. It is composed of the former Institute of Environmental Chemistry and the Ecology Center. RCEES' purpose is to promote the cooperation of environmental chemists, ecologists, and geoscientists to solve ecological and environmental issues at regional, national and global scales (RCEES 1991). Under the acting directorship of Huang Junxiong, RCEES has 516 staff.

With 34 laboratories within seven divisions, the size of RCEES and its interdisciplinary nature allow a wide variety of projects. Some of the activities are directly related to global change issues, others are generally related: (1) implementation of national and key environmental projects relevant to regional, urban, and rural economic development, such as acid rain in southwest China, Tianjin and Yichang environmental planning, and the Minjiang River Estuary environmental assessment; (2) participation in joint projects on global eco-environmental issues; (3) basic research in the fields of ecological chemistry, theoretical ecology, and carbohydrates chemistry; (4) research, development, and production of pollution control devices; and (5) provision of advisory and technical information services to central and local governments and enterprises.

The graduate education program is strong; 176 students have been enrolled. Two students have been awarded Ph.Ds, five others have completed their requirements for the degree, and 31 are doctoral students. Seventy-nine students have been awarded M.S. degrees, 24 have completed their requirements, and the remaining are in the process of completing their work for M.S. degrees (RCEES 1991).

RCEES has several well-equipped laboratories, with instrumentation that includes a high-resolution X-ray fluorescence spectrometer with double crystals, a high-frequency plasma emission spectrograph, a Fourier transform infrared spectrograph, a combined gas chromatograph and mass spectrometer, an X-ray diffraction and fluorescence spectrograph, a laser spectrometer, a UV/visible light spectrophotometer, an ion chromatograph, an atomic absorption spectrophotometer, a gas chromatograph, a high-pressure liquid chromatograph, a polarograph, a thermal energy analyzer, a differential thermal analyzer, and a computer work station. In addition, RCEES has a VAX-780 computer shared jointly with three other CAS institutes, direct computer connections with an institute in Germany, and about 20 PCs.

About one-third of RCEES's funding comes directly from CAS, the remainder from NEPA, NSFC, and contracts from local governments.

RCEES has a library with a collection of more than 110,000 volumes published in many different languages and a collection that comprises 349 Chinese and foreign scientific journals. In addition, a collection of about 15,000 documents from more than 100 institutions is kept in a regular information exchange link within the country. The Chinese unit of INFOTERRA (UNEP's international environmental database) is also based at RCEES. RCEES staff edit two important journals: *Environmental Chemistry* (*Huangjing Huaxue*, bimonthly, in Chinese) and *Environmental Science* (*Huangjing Kexue*, bimonthly, in Chinese). Additionally, Zhuang Yahui is the editor of *Acta Scientiae Circumstantiae* (*Huangjing Kexue Xuebao* [*Journal of Environmental Sciences*], quarterly, in Chinese), which is perhaps the top journal of environmental science being published in China.

## Research Highlights

Of the seven RCEES divisions, four are contributing to or have the potential to contribute to global change research. The Regional Eco-environmental Assessment and Planning Division has three ecology laboratories and a research group for agricultural and soil chemistry. The Eco-environmental Effects of Chemicals Division includes research groups for atmospheric chemistry, ecological chemistry, biogeochemistry, chemical ecology, organic chemistry, and ecotoxicology. The Environmental Analytical Chemistry Division has four laboratories. The Eco-environmental Information Studies Division has five research groups working on environmental databases and information.

RCEES has international programs with Germany, Japan (an acid rain study in Chongqing and at other sites in South China), and the United States. Atmospheric chemistry and air pollution studies on local and regional scales seem to be the most numerous of these collaborative programs.

Only a few of the programs at RCEES relate to global change. RCEES provided the following information on these programs as they relate to the panel's interests in (a) atmospheric chemistry and (b) atmosphere–land surface interactions. RCEES proposals for land cover change studies were not funded under the Eighth 5-Year Plan. The center will look for international collaboration, possibly with Australia, Germany, or the United States to fund this type of work. Currently, the only ongoing land use project is one to study land reclamation after mining operations.

### *Trace Gas Projects*

- Estimation of greenhouse gas emissions (Yang Wenxiang, PI)
- Studies of the fluxes of $N_2O$, $CH_4$, nonmethane hydrocarbons, and the concentration of carbon monoxide in northern China (planned IGAC contribution) (Su Weihan, PI)
- Simulation of trace gas emissions from agricultural biomass burning in China (Yang Wenxiang, PI)
- Study of clean coal application techniques and the control of $CO_2$ emissions (planned) (Zhao Dianwu, PI)
- Study of the effects of $CO_2$ and $O_3$ on the growth and production of rice, wheat, and corn
- Study of the effect of $CO_2$, $O_3$, $SO_2$ on the deciduous, broadleaved forest ecosystem[2]
- Modeling of climate change and agricultural ecosystems
- Study of forest ecosystems[3]
- Study of the role of emission and absorption of trace gases in forest ecosystems
- Study of trace gas emissions from soils
- Study of trace gas fluxes from different environmental systems
- Study of the chemical reaction of the hydroxide radical and oxygen with greenhouse gases
- Study of the vacuum ultraviolet photolysis of CFCs
- Estimation of the atmospheric residence time of some CFCs and $N_2O$ and the assessment of the impact of these gases on the $O_3$ layer (planned) (Yang Wenxiang, PI)
- Study of the key reactions of $O_3$ depletion

## Biogeochemistry
- Study and modeling of the biogeochemical cycles and of natural and anthropogenic emissions of carbon, nitrogen, and sulfur in China

## Precipitation Chemistry
- Research on the atmospheric chemical processes for the formation of acid rain in southwest China (Shen Ji and Zhao Dianwu, PIs)
- Study of the critical loadings of acid precipitation
- Cooperative study of a control policy for acid deposition
- Study of the impact of atmospheric deposition on forest ecosystems

In summary, RCEES is a dynamic institute with tremendous potential for contributing not only to the global change programs of China, but also in providing a bridge to U.S. programs. RCEES is in the process of adding a global dimension to their historical focus on local (urban air pollution) and regional (acid deposition) research. They have initiated a number of programs that address global-scale questions, especially in relation to the emission of radiatively active gases to the atmosphere and biosphere and atmospheric interactions (see projects listed above).

RCEES' strengths have been in the areas of local and regional issues of environmental chemistry, but continued development is limited by a lack of resources and facilities. As is common with other institutes, tremendous competition exists for funds and resources. This competition limits the development of new programs and limits the amount of intergroup and interinstitutional cooperation on questions relating to global change.

A number of scientists have excellent research records that address a wide variety of questions of interest to the international community. Concerning collaboration with U.S. scientists in the area of global change research, several RCEES scientists have established cooperative projects, for example, Su Weihan, Zhao Dianwu, and Liu Jingyi (who strongly supports international collaboration in her capacity as Secretary General of the United Nations Scientific Committee on Problems of the Environment-CAST program for China). Scientists are eager to build upon these collaborations.

## SHANGHAI INSTITUTE OF PLANT PHYSIOLOGY

The CAS Shanghai Institute of Plant Physiology was established in 1986 and is directed by Yang Xiongli. It is primarily a molecular and cellular biology institution and its main focus has shifted from physiology to tissue culture, biochemistry, and molecular genetics. As well as dealing with plants, research also has a strong microbiological thrust. The institute takes on 10 graduate students per year, and confers its own degree.

Given its current focus, the institute is not an obvious center for global change research. Yet, it has important assets and relevant projects, nevertheless. Wang Tianduo, who is working on a water use efficiency experiment, is a member of the pre-World War II school of scholars whose dignified excellence dots the Chinese academic landscape. Although he has been abroad, he has never been to the United States, where seven of his eight students now work on plant processes. Fu Wei and Ding Yang are two sharp graduate students with good modeling capabilities who are working under Wang.

## Research Highlights

Wang Tianduo's current research is centered around the development of a mechanistic model for atmosphere–plant linkages—photosynthesis, respiration, allocation and growth—water flux, and balance. This model, PGROW (which only has parameters for annual plants) is being applied chiefly to crops as part of the water use efficiency project on the Huang–Huai–Hai Plain in which Wang plays a leadership role (Chapter 4). This project has the potential to make important contributions to regional-scale global change research (Wang 1990). Very importantly, Wang is working actively on the problems of "scaling up" analyses. This experience and expertise of Wang and his students might be applied to the development of, or incorporation of China into, a general ecosystem model that links atmosphere–plant–soil through energy–water–chemical exchanges.

Two other projects should be noted for their relevance to global change research. First, Xu Daquan makes detailed gas exchange measurements and is interested in the direct effects of $CO_2$ enhancement on plants. He recently worked on this with Roger Gifford in Canberra, Australia and has some intriguing results on acclimation to higher $CO_2$. Xu Daquan and Qin Guoxiong are working on $CO_2$ enhancement field experiments by using natural $CO_2$ from a well north of the Yangtze River in Jiangsu Province. His laboratory is equipped with an oxygen sensor leaf chamber, an ADC infrared gas analyzer for lab/phytotron measurements, an ADC porometer/infrared gas analyzer for field measurements, and a Wisconsin-type phytotron (without $CO_2$ control facilities). Second, Yu Shuwen works on $SO_2$ and industrially-derived ethylene pollution effects on plants.

## SOUTH CHINA INSTITUTE OF BOTANY

The CAS South China Institute of Botany, located in Guangzhou under the directorship of Tu Mengzhao, has six departments—plant genetics, physiology, taxonomy, morphology, resources, and ecology—on a 16 ha campus. These departments provide good biological support for the ecologists and ecophysiologists who are the likely persons to be conducting global change research. The institute administers three ecological research stations and is responsible for the South China Botanical Garden, which was founded in 1958 and covers about 300 ha. The garden has currently more than 4,000 species, including some 500 species of medicinal plants. (Guangzhou has the main phytotaxonomic facilities in South China, including an extensive herbarium where about 700,000 specimens are stored.) The staff consists of 534 people, including 74 senior scientists, and approximately 25 M.S. students.

The institute has experimental rice paddies, greenhouses, and some environmental control chambers. Instruments include an antiquated transmission microscope, a fairly new scanning electron microscope, an organic compound mass spectrometer for the large volume of phytochemistry conducted, a LICOR porometer for plant leaf gas exchange measurements, an oxygen electrode for measuring photosynthesis and respiration of leaf samples, and a gas chromatograph with a flame ionization detector.

The institute's ecological research stations, two of which are in CERN (Chapter 4) are considered to be key to its participation in global change research. The Xiaoliang Artificial Tropical Forest Ecosystem Experiment Station, which is not in CERN, is far down the coast to the southwest and features a subtropical forest restoration project on heavily eroded soils, probably oxisols. Heshan Comprehensive Downland Experiment Station, which is in CERN, is mainly an agroecosystem station devoted to land reclamation (primarily through reforestation with exotic tree species), integrated agroforestry, animal husbandry, and aquaculture production (Yu and Wang 1990). Dinghushan Subtropical Forest Ecosystem Experiment Station[4] is located in an UNESCO Man and the Biosphere (MAB) reserve and is a CERN station. It features natural and semi-natural subtropical and tropical forest vegetation (up to 400 years old) extending over terrain nearly 1,000 m in elevation (CAS Undated, b). It appears that they have a thorough understanding of the natural histories of these locations, and Dinghushan, in particular, appears to offer excellent research potential.

The institute has no GIS or remote sensing capabilities, but researchers can contact Hu Suozhong at the Department of Geography at South China Normal University for remote sensing data. Because the Dinghushan station has been included in CERN, it should receive equipment and training for GIS. Computer facilities include eight PCs.

This institute has enjoyed three important cooperative programs with other nations that helped to introduce modern ecology and whet researchers' appetites for more. The first of these was with Germany in which a Cooperative Ecological Research Project was carried out with Hans Brunig of Hamburg University and others on ecosystem processes at the Xiaoliang station. The second is a MAB project on ecosystem restoration with Sandra Brown at the University of Illinois. The third is a MAB project headed by Orie Loucks at Miami University (Ohio) on the comparison of broadleaved forests. James Ehleringer of the University of Utah and Chris Field of Carnegie Institution at Stanford University have done collaborative ecophysiological work at this station.

Institute objectives for global change research revolve around monitoring processes at their research stations. Extrapolating local knowledge to a regional evaluation or developing methods for predicting change had not been considered. With increased knowledge of scientific principles, sampling techniques, modeling, and leadership, this institute could make a contribution by virtue of its geographical location, institutional facilities (including the stations), and the basic skills of the staff. One of its most important resources is the enthusiasm and eagerness of the staff for interaction and collaboration.

For the purposes of discussion, the following are examples of global change research the institute may wish to undertake:

- Describe land cover change for Guandong Province and southern China more generally over the last several decades (they have a 1970 vegetation map as a base).
- Develop models for predicting land cover change for specified scenarios of climate change (including crops).
- Design imaginative ways to measure the states of ecosystems at their stations by using biological measurements such as radial growth increment or degrees of herbivory.
- Devise ways to learn how subtropical vegetation will respond to elevated $CO_2$. (The institute has adequate photosynthetic measurement capacity.)
- Create linkages with other regional institutes to do regional appraisals.

## SOUTH CHINA SEA INSTITUTE OF OCEANOLOGY

The South China Sea Institute of Oceanology in Guangzhou, under the directorship of Sun Yuke, is one of two sister CAS institutes of oceanology, the other being the Qingdao Institute of Oceanology (see above) in Shandong Province. Within CAS, the Qingdao institute is responsible for the East China Sea and this one for the South China Sea. It confers the M.S. degree and has from five to 15 graduate students.

### Research Highlights

The institute is conducting mostly historical analyses that would be considered under the rubric of global change. Researchers are using a multisource (pollen, meteorological data, and archives) climate record for the last 2,000 years for Guangdong Province (compiled by Chen Shixun of Zhongshan University [Chen 1990]). Coral reef structures (Nie et al. 1991, Nie, Undated) are being studied. Nie Baofu has calibrated the lamina widths with seawater temperature over the last decade or so to estimate temperatures of the past several thousand years. Zhao Huanting described coral reef work based on 100 m of island coral for sea temperature proxy data. Zhu Yuanzhi is examining the physical rate of carbonate rock deposition as a function of temperature as another means of historical analysis. Zhang Qiaomin has done an historical study of the evolutionary geomorphology of a tidal inlet. His ongoing work is toward forecasting future change.

A large-scale, multi-institutional project has been ongoing for a number of years that addresses the rise in sea level versus change in land level due to tectonics for the entire China coast. This institute is charged with the South China Sea coastline on this project. As described by Chen Tegu, this project has historical significance, but as part of a forecasting program. At least three opinions on this phenomenon in China have been put forward. It is not clear, though, whether Ren Mei-e from Nanjing University is involved. Chen stated that they would welcome collaboration with the United States on this kind of work.

Lue Youlang is working on a 6 to 8 meter-long deep sea sediment core from the South China Sea. He says it goes back 230,000 years and he is doing oxygen-isotope work on it with Dr. Sacklette, a chemical oceanographer from the United Kingdom. A group including Lue was seeking funding for a similar core from Antarctica.

These are projects pertinent to PAGES, but some have prediction potential as well that may be relevant to LOICZ and JGOFS. More directly relevant to JGOFS is Chinese participation, specifically this institute, in the World Ocean Circulation Experiment (WOCE) and TOGA, and about which Gan Zijun had considerable knowledge. Blue water oceanography is evidently a rather active and important activity here. However, it is hard to say if it has higher priority than coastal zone work. This institute was originally founded for coastal zone work, but as they have accrued vessels (and bigger budgets), they have been able to expand into deep sea cruise work.

He Youhai described a physical oceanography project that is part of TOGA. One of the results of these cruises is measurement of sea surface temperature and currents in the South China Sea. These variables are definitely related to ENSO. One intriguing idea would be to model the relationship of ENSO with these variables and, with GCM scientists, their effect on the timing and strength of the summer monsoon to east China and typhoons.

On the coastal zone *per se*, a large, multidisciplinary, national-scale coastal zone program was conducted between 1980 and 1986. Individual coastal provinces were responsible for work on their shorelines. Unfortunately, no references or results were available. More generally, the Department of Estuarine and Coastal Studies, headed by Zhang Qiaomin, will be useful for research relevant to LOICZ.

## XI'AN LABORATORY OF LOESS AND QUATERNARY GEOLOGY

The Xi'an Laboratory of Loess and Quaternary Geology was established in 1984. Under the management of the Xi'an branch of CAS, the director is Liu Tungsheng, China's senior and leading scholar of geology, who also is a member of the PAGES Scientific Committee. Designated as an open laboratory in 1987, it employs a scientific staff of 10 and additional research fellows and students. This laboratory is one of the leading centers for research on past climate change, and an application has been made to be named a national key laboratory.

Its research program is organized around a central theme of documenting past climate change as recorded stratigraphically in the Loess Plateau from 2.5 million years ago to predict future climate trends and impacts. Researchers address this task by studying the properties, textures, and formation processes of loess and other Quaternary sediments. Loess in China covers 600,000 $km^2$; it is the largest loess deposit in the world. By virtue of its location on the Loess Plateau itself, the laboratory is near suitable geologic sites for carrying out field observations and experiments.

## Research Highlights

Scientists can carry out a broad range of geological, geochemical, and geophysical measurements that provide the basis for investigations of past climate change in North China, including geologic observations in the field and measurements by microscopy in the laboratory (An Zhisheng, Zhu Yizhi, and Zhou Jie, PIs), spore pollen analysis (Li Xiaoqiang, PI), radiocarbon dating (Zhou Weijian and Jiang Yu, PIs), stable isotope measurements (Liu Yu and Su Fuqin, PIs), thermoluminescence dating (Xie Jun, PI), magnetostratigraphy (Zheng Hongbo, PI), and chemical element analysis of present day dust aerosol (Zhang Xiaoye, PI).

Laboratory scientists have now established proxy sequences of paleoclimates and environments on time scales of 2.5 million years, 150,000 years, and 20,000 years B.P. for the Loess Plateau. Three of the major results are noteworthy:

- A theory has been put forth that paleo-Asian monsoon is a controlling factor in environmental changes in central China, for example, that variation in monsoon circulation may have caused variation in temperature, moisture, soil conditions, and plant growth during the past 20,000 years.
- A distributional model has been developed for paleo-environments around 18,000 years before present and at the Holocene optimum 9,000 to 5,000 years before present.
- An abrupt event about 10,000 to 11,000 years before present has been discovered when the summer monsoon weakened and the winter monsoon strengthened, corresponding to the Younger Dryas in the North Atlantic region.

Scientists have actively sought to establish collaboration with researchers at other institutions, including laboratories in the United Kingdom and the United States. An Zhisheng and Stephen Porter, University of Washington, are conducting a comparative study of the chronology and dynamics of Late Quaternary climate and environmental changes between central China and the northwestern United States. Based on research undertaken for the China and America Air–Sea Experiments (CHAASE), An Zhisheng, Zhang Xiaoye, and Richard Arimoto, University of Rhode Island, are conducting a comparative study of the atmospheric transport of soils by focusing on the interannual variability in the atmospheric dust concentrations and the meteorological conditions responsible for the concentration differences (Appendix C).

Results of the laboratory's research have been published in Chinese and foreign journals. One recent publication, in English, *Loess, Environment and Global Change*, edited by Liu Tungsheng (1991), is a timely and useful compilation of paleoclimate studies of loess in China. Notably, the papers presented in this volume demonstrate a wide diversity of institutions represented by collaborating authors and a solid degree of international collaboration with researchers from Europe and the United States.

## XINJIANG INSTITUTE OF BIOLOGY, PEDOLOGY, AND DESERT RESEARCH

The mission of the CAS Xinjiang Institute of Biology, Pedology, and Desert Research, under the leadership of Xia Xuncheng, is to study the arid environment of western Xinjiang Uighur Autonomous Region, including microclimate, soils, hydrology, vegetation, and land use. The institute is particularly concerned with the effects of land use change and climate change on desert and desert riparian vegetation; it is also concerned with water conservation and water management.

The institute administers five experiment stations, one of which, the Fukang Desert Ecosystem Observation and Experiment Station, was visited by panel members. This station is a new and fairly well-equipped facility near the desert margin. The laboratory is equipped for basic soil physical and chemical measurements, and maintains a complement of basic microclimate stations in various desert environments. The station is new and facilities are still being developed. In contrast to the instrumentation of the atmospheric institutes, this ecological station had relatively less expertise in climatological measurements, and future collaboration could aid in enhancing the quality and comprehensiveness of microclimate studies.

Research focuses more on local ecological and environmental problems than on large-scale changes, however it is documenting climate change and ecological change. Scientists study a transect from desert at the southern edge of the Junggar Basin to a glacial lake (*Tianchi*, a MAB reserve) in the nearby montane region of the Tianshan Mountains. This transect provides an interesting cross section of an area influenced strongly by climate variability and change and by intense human land use. Opportunities for collaboration on both ecological and human dimensions of global change are excellent. An additional opportunity to develop an even longer desert–mountain transect exists between the Fukang station and the CAS Lanzhou Institute of Glaciology and Geocryology's Tianshan Glaciology Research Station. Despite this opportunity, global change research is a new priority for the station research group, and, consequently, it is not a well-developed area of strength.

Personnel seemed enthusiastic and knowledgeable about their region. The training of the more senior scientists is fairly conventional (disciplinary) but these individuals seemed open-minded and broad in their interests. Younger scientists included a sprinkling of scientists with strong interests in plant physiology and biogeochemistry.

This institute could be a good partner in potentially excellent research documenting the interaction of climate variability (clearly documented changes in regional hydrology have occurred over the past decade in the Tianshan Mountains and their watersheds) and human land use. Collaboration with the U.S. Long-Term Ecological Research program could be particularly fruitful, given the existence of several CERN activities in the region. The documented reductions in rain and snowfall over the 1980s in this region permit use of (possibly) natural climate variability to probe the response of arid mountain–desert landscapes to climate change.

## ZHONGSHAN UNIVERSITY

Although panel members were not able to visit Zhongshan University, research from that university was described during a panel member's visit to the CAS South China Sea Institute of Oceanology.

Chen Shixun, a circulation modeler from the Department of Atmospheric Sciences, has worked with someone at Colorado State University and his work is rather limited due to insufficient computing power. He does not interact with the researchers at the CAS Institute of Atmospheric Physics. Nevertheless, he presented some interesting modeling work that linked climate features with winds and shoreline geomorphology, which would be pertinent to LOICZ.

## NOTES

1. AVHRR data are received by the State Meteorological Administration Satellite Center from stations in Beijing and Guangzhou. A third station in Urumqi in western China has not been operating for some time. Peking University's Institute for Remote Sensing Technology and Application has a ground station that also receives AVHRR data.

2. The project is based on the long-term impact of climate and air pollution on the forest in Mantounguog Forest Observation Station.

3. The objective is to provide specific data on China's forest ecology for input to global change studies. The methodology includes soils and plant classification and their mapping, experimental studies of material and energy flows, measurements of atmospheric gas concentrations ($CO_2$, nitrogen oxides, $SO_2$, and $O_3$), and the comparison of differences between atmosphere in urban Beijing and nearby mountain atmo-

sphere. Comparisons will be made in different seasons. Trends will be monitored for 5 years and will include some computer modeling. Mountain area studies will be conducted at the CAS Beijing Forest Ecosystem Station, which is located about 2 hours west of Beijing. The cost of the project is 1.8 million *yuan*, and funding is from SSTC and NSFC (Feng Zongwei and Zhuang Yahui, PIs).

4. *Tropical and Subtropical Ecosystems* is edited at the Dinghushan station. Volumes include papers about Dinghushan. It is published in Chinese with English abstracts and graphs.

# B

# Global Change Projects Listed by the National Natural Science Foundation of China[1]

## INTERNATIONAL GLOBAL ATMOSPHERIC CHEMISTRY PROJECT

### General Projects[2]

- Study of the variations of the atmospheric ozone ($O_3$) layer, Wei Dingwen, principal investigator (PI), Chinese Academy of Sciences (CAS) Institute of Atmospheric Physics (1990–1992)
- Remote sensing and numerical simulation of atmospheric $O_3$, Wang Guiqin, PI, Peking University (1990–1991)
- Study of the measurement of stratospheric $O_3$ distribution with LIDAR, Hu Huanling, PI, CAS Anhui Institute of Optics and Fine Mechanics (1990–1992)
- Study and simulation of the interaction of $O_3$ and general circulation, Chen Yuejuan, PI, CAS University of Science and Technology of China (1991–1993)
- Study of methane ($CH_4$) flux in a rice-growing area with vertical gradient method, Ki Xingsheng, PI, Chinese Academy of Meteorological Sciences (CAMS) (1989–1990)
- Analysis of the background data of Chinese dust storm formation and transportation, Ma Peimin, PI, CAMS (1989–1990)
- Preliminary study of the impact of increased carbon dioxide ($CO_2$) on grain production in China, Deng Genyun, PI, Beijing

Municipal Academy of Agricultural and Forestry Sciences (1989–1991)
- Influence of Chinese vegetation on the concentration of $CO_2$ in the atmosphere, Xu Songling, PI, Research Center for Eco-Environmental Sciences (RCEES) (1991–1993)
- Model of $CO_2$ exchange on various forest ecotypes in the southern subtropical zone, Sun Guchou, PI, CAS South China Institute of Botany (1990–1992)

A program of cooperative research on atmospheric chemistry between China and the United States consists of four subjects: (1) site selection for a continental baseline station in China through joint air sampling in order to establish a continental baseline station for long-term monitoring of gases and aerosols on the Qinghai–Tibet Plateau; (2) joint planning and participation in the Global Tropospheric Experiment Pacific Exploratory Mission in order to investigate the atmospheric budgets and photochemistry of trace gases and aerosols in the atmosphere over China and over the Pacific; (3) measurement of greenhouse gas fluxes in China, such as $CH_4$, nitrogen oxides, and other radiatively or chemically important species; and (4) development and application of regional and global models of atmospheric chemistry and research on related effects on environment and climate.

## PAST GLOBAL CHANGES PROJECTS

### General Projects

- What changes have already occurred: the scales and scaling law of global change, Liu Shida, PI, Peking University (1992–1993)
- Study of the interannual variability of the East Asian summer monsoon and the annual prediction of rainfall in China and typhoon frequency, Chen Lieting, PI, CAS Institute of Atmospheric Physics (1987–1989)
- Recent trends and regional differentiation of snow variation in western China, Li Peiji, PI, CAS Lanzhou Institute of Glaciology and Geocryology (1989–1991)
- Study of dynamic change in the environment of Nanjing and western Zhejiang based on contents of chemical elements in tree rings, Qian Junlong, PI, CAS Nanjing Institute of Geography and Limnology (1991–1993)
- Study of the historical evolution of life-supporting environment in China during the past 2,000 years.

- Chronological table of climate disasters in Chinese history, Zhou Yinlin, PI, Fudan University (1991–1992)
- Establishment of a climate series for the last 2,000 years in south China, Chen Shixun, PI, Zhongshan University (1987–1989)
- Climate–environmental records covering the last 2,000 years by using an ice core from the Dunde ice cap in the Qilianshan Mountains, Xiao Lingmu, PI, CAS Lanzhou Institute of Glaciology and Geocryology (1987–1989)
- Evolution of agricultural regions and the agroecological environment in Yellow River Valley, Shi Nianhai, PI, Shaanxi Normal University (1991–1993)
- Geochemical study of sedimentary records during the past 200 years and on setting particles [sic] at all seasons in plateau lakes, Wan Guojiang, PI, CAS Guiyang Institute of Geochemistry (1991–1993)
- Study of indicative factors for ecological environments of red corals (*Corallium*), Zou Renlin, PI, CAS South China Sea Institute of Oceanology (1991–1992)
- Exploration of water temperature variability low of South China Sea [sic] in the past 100 years by means of reef coral growth rate studies, Nie Baofu, PI, CAS South China Sea Institute of Oceanology (1991–1993)
- Quantitative study of the synoptic-climate characteristics and reconstruction of climate series for the last 1,000 years in China, Zhang De'er, PI, CAMS (1991–1993)
- Study of the historical evolution of the life-supporting environment in China during the glacial–interglacial cycles in the Late Quaternary Period
- Evolution of Quaternary glaciers and climate change in the western Kunlunshan Mountains, Zheng Benxing, PI, CAS Lanzhou Institute of Glaciology and Geocryology (1988–1990)
- Climate and environmental study of ice cores from Tibet, Yao Tandong, PI, CAS Lanzhou Institute of Glaciology and Geocryology (1990–1992)
- Environmental evolution in forest and grassland areas of northeast China during the last 10,000 or 20,000 years, Sun Xiangjun, PI, CAS Institute of Botany (1987–1990)
- Environmental evolution of the transitional agriculture–animal husbandry zone in north China in the Holocene Epoch, Zou Tingru, PI, Geography Department, Beijing Normal University (1988–1990)
- Changes in the Holocene environment and the evolution of

human civilizations in northern and western China, Ying Jonsheng, PI, CAS Institute of Geography (1988–1990)
- Environmental evolution during the Holocene Epoch around Daqingshan, Inner Mongolia Autonomous Region, Cui Haiting, PI, Peking University (1988–1991)
- Modern changes and measures of the ancient geographic environment in northern Xinjiang Uighur Autonomous Region, Han Shuti, PI, Geography Department, Xinjiang University (1988–1990)
- Late Pleistocene environmental history in the east central area of the Qinghai–Tibet Plateau, Liu Tungsheng, PI, CAS Institute of Geology (1988–1990)
- Paleoclimatology and chronology of the Loess Plateau, An Zhisheng, PI, CAS Xi'an Laboratory of Loess and Quaternary Geology (1988–1990)
- Migration of the transitional zone between desert and loess areas in Erdosi and its climate changes during the last 1,500 years, An Zhisheng, PI, CAS Xi'an Laboratory of Loess Quaternary Geology (1990–1992)
- Study of paleo-oceanography in the western Pacific Ocean since 150,000 years before present, Cang Shuxi, PI, CAS Qingdao Institute of Oceanology (1988–1990)
- Comparative study of Quaternary paleo-oceanographic events in the South China Sea and its application, Ye Zhizheng, PI, Institute of Oceanographic Geology, Ministry of Geology and Mineral Resources (1989–1991)
- Climate change during the last deglaciation: records in the China Sea, Wang Pinxian, PI, Tongji University (1991–1992)
- Study of desertization and derivate [sic] sediments of the Yellow Sea during the last stage of Late Pleistocene Epoch, Zhao Songling, PI, CAS Qingdao Institute of Oceanology (1991–1993)
- Study of paleomagnetism and their correlation [sic] from Speleothems collected in China, Liu Yuyan, PI, China University of Geosciences (Wuhan) (1991–1993)
- Study of the secular variation on magnetic field effects on paleoclimate change, Su Weimin, PI, CAS Qingdao Institute of Oceanology (1991–1994)
- Paleolimnological study of the limestone district in central Yunnan Province, Song Xueliang, PI, Yunnan Institute of Geological Science (1987–1988)
- Study of environment changes in the Holocene Epoch and the existing environment in the future [sic] on the Nuo Ergai Plateau, Wang Fubao, PI, Nanjing University (1991–1992)

- Study of paleoclimate by isotopic dating and temperature determination in the past 100,000 years in Fujian, Hing Ashi, PI, State Oceanographic Administration (SOA) Third Institute of Oceanography (1990–1992)

### Key Projects

- Evolution of natural environment in the northeast Qinghai–Tibet Plateau during the Late Quaternary Period

### Major Projects

- Study of the changing trends in the life-supporting environment in the coming 20 to 50 years in China (a national key project), Ye Duzheng, CAS Institute of Atmospheric Physics (1991–1995)
- Dynamic processes and environmental changes since 15,000 before present along a corridor from Xinjiang Uighur Autonomous Region to the Yellow Sea Continental Shelf

## BIOSPHERIC ASPECTS OF THE HYDROLOGICAL CYCLE PROJECTS

### Major Projects

- Investigation of land–atmosphere interaction in Heihe Region, Gansu Province (HEIFE) (1987–1994)
- Study of water-saving agriculture on the North China Plain (1992–)

## LAND-OCEAN INTERACTIONS IN THE COASTAL ZONE PROJECTS

### General Projects

- Research on the relation between sea-level change and near sea continental motion, Zhang Chijun, PI, CAS Institute of Geodesy and Geophysics (1991–1993)
- Effects of sea-level change on the environment and economy in the Pearl River Delta, Huang Zhenguo, PI, Guangzhou Institute of Geography (1989–1990)
- Study of a geographical database with shoreline as a baseline, Zhang Jin, PI, CAS Institute of Geography (1989–1990)

- Environmental evolution of the Guangdong coastal region since the Middle Holocene Epoch and its influence on development, Li Pingri, PI, Guangzhou Institute of Geography (1987–1989)
- Recent sea-level changes in China and their effects on the deltas of the Yellow, Yangtze, and Pearl Rivers, Ren Mei-e, PI, Nanjing University (1990–1992)
- Comparisons of environmental change in coastal areas of South China with those of California during the Middle and Late Holocene Epoch, Li Pingri, PI, Guangzhou Institute of Geography (1990–1992)
- Changes of sea level and their influence on the Yangtze River Delta, Wang Baocan, PI, East China Normal University (1987–1989)

## JOINT GLOBAL OCEAN FLUX STUDY PROJECTS

### General Projects

- Observations and statistical analysis of the oceanic whitecap coverage, Chen Junchang, PI, CAS South China Sea Institute of Oceanology (1991–1993)
- Moist flux over the sea in high winds, Qian Zhengxu, PI, CAS Qingdao Institute of Oceanology (1987–1989)
- Wave dynamics in the microwave spectrum range and the strong exchange on the sea surface, Yuan Yeli, PI, SOA First Institute of Oceanography (1988–1990)
- Study of dynamic mechanics of the Kuroshio and atmosphere circulation and their interrelation, Qin Zenghao, PI, Qingdao Ocean University (1989–1991)

### Key Projects

- Marine flux in a typical area within the Continental Shelf of China (pending)

### Major Projects

- Preliminary study of sea-level and climate changes in China and their trends and impacts, Shi Yafeng, PI, CAS Nanjing Institute of Geography and Limnology (1987–1991)

## HUMAN DIMENSIONS OF GLOBAL ENVIRONMENTAL CHANGE PROGRAM PROJECTS

### General Projects

- Comprehensive survey of China's population, resources, society, environment, and ecology, Ma Bin, PI, Institute 701, Ministry of Aviation and Spaceflight (1990–1993)
- Study of the mechanism of transforming waterlogged land to three-dimensional dike–pond system, Zhong Gongfu, PI, Guangzhou Institute of Geography (1989–1991)
- Economic development and environment evolution law [sic] oases, Huang Shengzhang, PI, CAS Institute of Geography (1988–1990)
- Research on the structure, function, and optimization of the natural-social-economic compound system of the Beijing–Tianjin region, He Shanyu, PI, CAS Institute of Automation (1988–1990)
- Research on the optimization models of economic development and environmental capacity in the Urumqi region, Xie Xiangfang, PI, CAS Xinjiang Institute of Geography (1989–1990)
- Research on the situation of global natural resources and condition analysis [sic] of China, Li Wenhua, PI, Commission for Integrated Survey of Natural Resources (CISNAR) (1990–1992)
- Research on water resource utilization structure and models in the Bohai coastal zone (including Liaodong and Shangdong Peninsulas), Cheng Tianwen, PI, CAS Institute of Geography, (1990–1992)
- Mechanisms of controlling and coordinating of human–land system and patterns of regional development, Mao Handing, PI, CAS Institute of Geography (1991–1993)

### NOTES

1. From a paper by Zhang Zhifei, "Strengthening Coordination for More Effective Funding for IGBP," presented at the IGBP Regional Meeting for Asia, February 1991. [This list is not comprehensive of all global change research being undertaken nor of the research described in this report.]

2. NSFC projects are divided into three categories: general, key, and major. General projects receive relatively small grants, and awards deliberately support the widest range possible of ideas, people, and geographical location. General projects usually receive between 60,000 ($11,000) and 70,000 ($12,700) *yuan* over 3 years. Key projects are selected from the general pool of applications for their academic significance and for their potential applications. Key project funding is usually between 500,000 ($91,000) and 700,000 ($127,000) *yuan* for 3 years. Major projects are comprehensive topics selected for their importance to the development of science and to Chinese economic development goals. Funding levels begin at 2 million *yuan* ($364,000) for 5 years.

# C

# Selected Bilateral and Multilateral Global Change Projects

# BILATERAL PROJECTS UNDER THE U.S. SCIENCE AND TECHNOLOGY UMBRELLA AGREEMENT

| Protocol | Program | Lead U.S. Agency | Lead Chinese Agency | Dates | Objectives |
|---|---|---|---|---|---|
| Agriculture | Agricultural Environment Protection | USDA | MOA | 5/91 | Study of sustained ecological agriculture technology; equipment and techniques for agricultural environment; techniques for monitoring greenhouse gas emissions from rice paddies, livestock, and poultry raising; disposal and comprehensive use of livestock and poultry manure. Data are available. Contact: Lucia Claster, USDA Office of International Cooperation, (202) 690-2867; Chinese PI: Zhang Wenqing, MOA. |
| Agriculture | Agricultural Productivity | USDA | MOA | 6/91, 5/92 | Study of agricultural productivity in arid and semi-arid regions and water and soil erosion control techniques. In 1991, evaluated new plant species and crop and livestock production systems in Gansu Province and the Xinjiang Uighur Autonomous Region. Examined desert reclamation, stabilization, and conservation activities. Trip report is available. Contact: Lucia Claster, USDA, (202) 690-2867; U.S. PI: Robert Lansford, New Mexico State University; Chinese contact: Liu Congmeng, MOA. |
| Agriculture | Methane Emissions | USDA | MOA | 7/92 | Monitoring of rice paddy emissions and the control of methane emissions from dairy cattle. Contact: Lucia Claster, USDA Office of International Cooperation, (202) 690-2867. |
| Agriculture | Renewable Energy | USDA | MOA | 9/92 | Study of techniques of turning agricultural wastes (crop residue, straw, manure, grain hulls, and wood scraps) into feed, complex fertilizers, or electricity; disposal of animal and poultry waste; technology for the commercialization of solar, wind, and light power electricity generators. Contact: Lucia Claster, USDA Office of International Cooperation, (202) 690-2867. |

| | | | | |
|---|---|---|---|---|
| Agriculture | Utilization of Water | USDA | MOWR | 10/92 | Study of surface water, soil water, and ground water in areas in need of water resources or that are liable to drought, waterlogging, or alkalization. Contact: Lucia Claster, USDA Office of International Cooperation, (202) 690-2867. |
| Atmosphere | Atmospheric Chemistry Modeling | NASA | CAMS | 1990- | Cooperation via joint modeling workshops. The first workshop was held in Shanghai in 1990, where papers were presented on the budget and chemistry of trace gases, acid deposition modeling, chemistry–climate interactions, and stratospheric modeling. A second workshop was proposed for April 1992, which it is hoped will develop specific recommendations for substantive areas of cooperation in modeling. Contact: Robert McNeal, NASA, (202) 453-1479, U.S. PI; Chinese PI: Zhou Xiuji, CAMS. |
| Atmosphere | PEM-West | NASA | CAMS | 1991- | Study of anthropogenic impacts on the biogeochemical cycles of carbon, nitrogen, ozone, sulfur, and aerosols due to long-range transport of air pollutants from Asia and North America. The first round of experiments was completed in 1991 (data sets not yet available). Another mission is expected to take place in 1994. NOAA and CAMS run an intensive ground station in the PEM-West network capable of measuring many particulate species. Contact: Shaw Liu, NOAA, (303) 497-3356; U.S. PI: Robert McNeal, NASA (202) 453-1479; Chinese PI: Zhou Xiuji, CAMS, 86-1-832-7390. |
| Atmosphere | Continental Baseline Monitoring Station | NOAA | SMA | 1990- | Establishment of a baseline monitoring station in Qinghai Province for the collection of weekly air samples of carbon dioxide, carbon monoxide, and methane. Qinghai is a unique clean-air, inland station, and the data will be used as a base for long-term atmospheric chemistry measurements. Contact: James Peterson, NOAA, (303) 497-6074, U.S. PI; Chinese PI: Zhou Xiuji, CAMS. |

BILATERAL PROJECTS—*continued*

| Protocol | Program | Lead U.S. Agency | Lead Chinese Agency | Dates | Objectives |
|---|---|---|---|---|---|
| Atmosphere | Monsoon Program | NSF | CAMS | 1983- | Causes and effects of the Asian monsoon and its connection with weather patterns in other parts of the world, including studies of global teleconnections of the winter monsoon, heat and moisture budgets of monsoon convection, monsoon circulations and associated cloud systems, mesoscale dynamics and the structure of mesoscale convective systems, and the response of the tropical atmospheric circulation to monsoon heating. Rainfall data sets available from the Chinese side. Contact: Pam Stevens, NSF, (202) 357-9887; U.S. PI: K.M. Lau, NASA; Chinese PI: Chen Longxun, CAMS. |
| Atmosphere | Climate Studies | NSF | CAMS | ongoing | Collaborative research projects on the climatic effects of greenhouse gases and aerosols. Contact: Jay Fein, NSF (202) 357-9892; U.S. PI: Wei-Chyung Wang, State University of New York, Albany, (518) 442-3357; Chinese PI: Ding Yihui, CAMS. |
| Atmosphere | Tibet Plateau and Mountain Meteorology Experiment | NSF | SMA | 1985-1987 | Study of how large terrain complexes influence atmospheric heating and subsequent air motions. Measurements of radiation, heat and moisture fluxes near the ground to gain insight into the meteorological effects of the Tibet Plateau. Data sets available (preliminary data in the *Bulletin of the American Meteorological Society*, 1987, Vol. 68, pp. 607-615). Contact: Ron Taylor, NSF (202) 357-7624; U.S. PI: Elmar Reiter, Colorado State University, (303) 442-2200; Chinese PI: Zheng Qinglin, SMA. |

| | | | | | |
|---|---|---|---|---|---|
| Atmosphere | Tropical Ocean Global Atmosphere (TOGA) | NSF | SMA | ongoing | Although this topic is primarily a NOAA Marine and Fishery Protocol issue, a meteorological component to model the coupled ocean–atmosphere climate system (part of COARE) and ENSO prediction comes under the Atmospheric Protocol. Contact: Jay Fein, NSF, (202) 357-9892; U.S. PI: Mark Cane, Columbia Lamont-Doherty Geological Observatory, (914) 359-2900 x344; Chinese PI: Zhou Xiaoping, SMA. |
| Atmosphere | Climate Reconstructions | NSF | SMA | ongoing | Historical climate data extraction and tree-ring core collection and analysis to yield climatic records extending back at least 300 years. In 1991, discussions were held regarding the development of an exchange program for historical research including ice-core analyses and the construction of historical climate data comparison models. A joint paper by Thomas Crowley (Applied Research Corporation, [409] 846-1403) and Zhang Jiacheng (SMA) on the reconstruction from historical and paleorecords of Chinese climate variability and change from 1470 to 1979 was completed and published in the *Journal of Climate*. Contact: Jay Fein, NSF, (202) 357-9892; U.S. PIs: Pao Wang, University of Wisconsin, (608) 262-6479, and Lisa Graumlich, University of Arizona; Chinese PI: Ding Yihui, CAMS. |
| Environment | CFC-Substitutes for Household Refrigerators | EPA | NEPA | 4/91- | Cooperative projects to test short-term methods for reducing CFC-11 and CFC-12 use in Chinese refrigerators and to test commonly available alternative designs that use long-term non-ozone depleting, energy-efficient substitutes for CFCs in refrigerators. The Chinese Ministry of Light Industry is also involved in this project. Contact: Stephen Andersen, Global Change Division, EPA (202) 233-9069. |
| Environment | Elimination of Halons | EPA | NEPA | 1/92- | Project to eliminate unnecessary halons in order to protect the stratospheric ozone layer and to improve fire safety in China. Coordination of cooperative projects to recall, recycle, and store halons from non-essential applications; to replace halon |

BILATERAL PROJECTS—continued

| Protocol | Program | Lead U.S. Agency | Lead Chinese Agency | Dates | Objectives |
|---|---|---|---|---|---|
| | | | | | fire extinguishers and systems with modern dry chemicals, carbon dioxide, foam, and water fire extinguishers; and to convert halon chemical factories to the manufacture of other products. The Chinese Ministry of Public Safety is also involved in this project. Contact: Stephen Andersen, Global Change Division, EPA, (202) 233-9069. |
| Environment | Energy Efficient Refrigerators | EPA | NEPA | | Under the auspices of the CFC Substitutes for Household Refrigerators project, this is an effort to increase the efficiency and decrease the energy consumption of Chinese refrigerators. The project is implemented through the Beijing Household Electrical Appliance Research Institute of the Ministry of Light Industry, whose scientists are researching substitutes for CFCs to increase the efficiency of refrigerators. Contact: John Hoffman, Global Change Division, EPA, (202) 233-9190. |
| Environment | Ozone Protection Projects | EPA | NEPA | 12/90- | Based on the Cooperation Principles for Cooperative Ozone Protection Projects. Joint research on the relationship between CFC and halon emissions and ozone depletion; the effects of ozone depletion; and CFC and halon substitutes and alternatives. Contact: Stephen Andersen, Global Change Division, EPA (202) 233-9069. |
| Environment | Rice Paddy Methane | EPA | NEPA | | The focus is on two areas: (1) methane analyses through soil survey and the monitoring of methane emission coefficients and (2) comparative analyses of methane levels under different soil fertility, irrigation, and cultivation practices. Within NEPA, the work is conducted at the Chinese Research Academy for |

| | | | | |
|---|---|---|---|---|
| Environment | Air Pollution Transport and Transformation | EPA | NEPA | 1988- | Studies of the dispersion of pollutants in the atmosphere; transformation of $SO_2$ to $SO_3$ in the atmosphere; rules of pollutant diffusion in areas of complex topography (1988-89); atmospheric chemical processes and models (ongoing since 1988); the transformation of harmful substances contained in soot particles (ongoing since 1988); mountain cloud chemistry project (1989-91); and cooperation on the Global Trends Network (1979-89). NOAA and CAS are also involved in this program. Contact: William Wilson, EPA (Research Triangle Park), (919) 541-2551. |
| Environment | Coal-bed Methane | EPA | MOE | 4/90- | The project is aimed at reducing atmospheric methane emissions, expanding the recovery and use of coal-bed methane, and providing substantial quantities of clean-burning natural gas. Current work is focused on resource assessment and the identification of demonstration project sites. The project is developing pilot projects in areas with documented coal-bed methane potential and creating a program to train Chinese engineers. No data are available. NEPA and SMA are also involved in this project. Contact: Dina Kruger, Global Change Division, EPA, (202) 233-9039; Chinese PI: Li Xuecheng, Ministry of Energy. |
| Fossil Energy | Analysis of General Circulation Models | DOE | CAS | 8/87- | Task One of the DOE-CAS Joint Research on the Greenhouse Effect. Analysis of various current climate models to improve general circulation models (GCM), specifically, the development of the CAS Institute of Atmospheric Physics' (IAP) two-level GCM. The *Bulletin of the American Meteorological Society* 73(5) 1992 describes this joint project. Data sets are available. Contact: T.K. Lau, DOE, (202) 586-9249; U.S. PI: R. Cess, University of New York at Stony Brook, (516) 632-8321; Chinese PI: Zeng Qingcun, Institute of Atmospheric Physics. |

Environmental Sciences and the Nanjing Institute of Environmental Science. Contact: Dennis Tirpak, Climate Change Division, EPA, (202) 260-8825.

## BILATERAL PROJECTS—continued

| Protocol | Program | Lead U.S. Agency | Lead Chinese Agency | Dates | Objectives |
|---|---|---|---|---|---|
| Fossil Energy | Preparation Paleo-, Historical, and Instrumental Climate Data | DOE | CAS | 8/87- | Task Two of the DOE-CAS Joint Research on the Greenhouse Effect. Compilation of a climate database which will be used to estimate natural climate variation, to study climate changes, and to test climate models. Chinese historical climate information will be analyzed to establish relationships between physical processes and climate changes, with special attention to the case studies of wet and dry periods, desertification, and increased atmospheric carbon dioxide. Data sets are available. Contact: Michael Riches, DOE, (301) 903-3264; U.S. PIs: Sultan Hameed, State University of New York (SUNY), Stony Brook, (516) 632-8319, Wei-Chyung Wang, SUNY, Albany, (518) 442-3357; Chinese PIs: Zhang Peiyuan, Institute of Geography, Fu Congbin and Huang Runhui, Institute of Atmospheric Physics. |
| Fossil Energy | Relationship Between Large-Scale and Regional-Scale Climate | DOE | CAS | 8/87- | Task Three of the DOE-CAS Joint Research on the Greenhouse Effect. Definition of temporal and spatial characteristics of climate regions by using regional climate data from a project database, and to estimate how such characteristics may be affected by global warming. Studies have analyzed desertification in northern China and its relationship to precipitation fluctuations and the *El Niño* Southern Oscillation. Data sets are available. Contact: Michael Riches, DOE, (301) 903-3264; U.S. PIs: Sultan Hameed, SUNY, Stony Brook, (516) 632-8319, Wei-Chyung Wang, SUNY, Albany, (518) 442-3357; Chinese PIs: Zhang Peiyuan, Institute of Geography, Fu Congbin and Huang Runhui, Institute of Atmospheric Physics. |

| | | | | |
|---|---|---|---|---|
| Fossil Energy | Atmospheric Measurements of Methane | DOE | CAS | 8/87- | Task Four of the DOE-CAS Joint Research on the Greenhouse Effect. Production of data on methane fluxes from various sources in China, particularly from rice paddies and biogas pits. The study tries to determine the concentrations and trends of various greenhouse gases at rural and urban continental stations. The study will evaluate the role of Chinese methane emissions in the global methane cycle. Data sets are available. Contact: Michael Riches DOE, (301) 903-3264; U.S. PIs: Rei Rasmussen and M.A. Khalil, Oregon Graduate Institute of Science and Technology, (503) 690-1077; Chinese PI: Wang Mingxing, Institute of Atmospheric Physics. |
| Marine/Fishery | TOGA/COARE | NOAA | SOA | 1985- | Study of the effect of ocean heat transport on climate variability by measuring ocean circulation and mixing, atmospheric convection and the fluxes of heat, momentum, and moisture. Research will improve forecasting of long-term climate change. Because tropical convection and latent heat release are among the main driving forces for atmospheric circulation in higher latitudes, study of this area is key to understanding the mechanisms controlling the *El Niño* Southern Oscillation phenomenon. NSF and NASA are also involved on the U.S. side and MOA and SMA on the Chinese side. Contact: James Bizer, NOAA, (301) 427-2089 x24. |
| Marine/Fishery | Oceanographic Data and Information Cooperation | NOAA | SOA | 1985- | The U.S. National Oceanographic Data Center and the SOA National Marine Data and Information Service exchange oceanographic data (hydrographic, model observations, etc.), atmospheric observations, buoy data, current measurements, and sea level studies that are relevant to global change research. Contact: Ron Moffatt, NOAA, (202) 606-4571. |

## Summaries of Selected University-Level Bilateral Projects

| Program | Lead U.S. Agency | Lead Chinese Agency | Dates | Objectives |
|---|---|---|---|---|
| Biogeochemical Cycling of Atmospheric Trace Elements and Mineral Aerosol Over Central and Eastern Asia | URI | CAS | 6/91- | Experiments focus on the atmospheric transport of soils, in particular, on the interannual variability in atmospheric dust concentrations and the meteorological conditions responsible for the concentration differences. A second focus will be to evaluate the relationships between contemporary mineral aerosol particles and loess/paleosol sequences in China. U.S. PI: Richard Arimoto, University of Rhode Island, (401) 792-6235; Chinese PI: An Zhisheng, Xi'an Laboratory of Loess and Quaternary Geology. |
| Chinese Ecological Research Network (CERN) | UNM | CAS | 1992- | Research data management training course to improve the management of CAS ecological data. U.S. PIs: James Gosz and James Brunt, University of New Mexico, (505) 277-9342; Chinese PI: Zhao Jianping, Bureau of Resources and Environmental Sciences, CAS. |
| Chronology and Dynamics of Late Quaternary Climatic and Environmental Changes | UWA | CAS | 1991- | Research compares paleoclimatic indices obtained from detailed environmental records in China and in the U.S. Pacific Northwest with results of paleoclimatic modeling experiments that produce values for former wind, temperature, and precipitation at specific times in the past. U.S. PI: Stephen Porter, University of Washington, (206) 543-1166; Chinese PI: An Zhisheng, Xi'an Laboratory of Loess and Quaternary Geology. |
| Climate Change Impacts | Climate Institute | CAS | 8/91- | The Climate Institute is a policy organization promoting links between CAS and international scientists and an understanding of global change, especially impact assessments and GCM modeling and training. Coordinates joint meetings between China and international experts, e.g., |

| | | | |
|---|---|---|---|
| | | | the Symposium on Climate Change Impacts in Beijing in 1991, with a follow-up meeting planned for 1993 Beijing) to discuss and analyze impacts and response strategies. The institute's work with China involves cooperation with Dick Ball of DOE, Steve Leatherman (United States) and Nobuo Minura (Japan) who study of the vulnerability of the Chinese coast, and CSIRO (Commonwealth Scientific Industrial Research Organization) of Australia. Contact: Ata Qureshi, Climate Institute, (202) 547-0104. |
| Global Change and Terrestrial Ecosystems | UM, UMD | CAS | | International collaborations with laboratories at the Universities of Miami and Maryland on climate–vegetation interactions. Chinese PI: Zhang Xinshi, Institute of Botany; U.S. PIs: Mark Harwell, University of Miami, (305) 361-4157, Alan Robock, University of Maryland, (301) 454-5089. |
| Global Trends Network | NOAA | NEPA | 1979- | Measurement of precipitation composition in remote areas, to be used as a baseline in a study of change in regions with high levels of human activity, especially in the United States and China. This network was established in 1979 as part of the Global Precipitation Chemistry Program. The China station was established in 1987 under the U.S.-Chinese Environment Protocol, but since 1989, the program has been administered by NOAA and the University of Virginia. Contact: Rick Arts, NOAA, (301) 713-0295; U.S. PI: James N. Galloway, University of Virginia, (804) 924-0569. |
| Man and the Biosphere | National Committee for MAB | National Committee for MAB | 1987-1991 | Measurement models and training in the areas of wood decomposition, nutrient cycling, species, replacement, and nitrogen cycling. The South China Institute of Botany has two cooperative projects, an ecosystem restoration project (Sandra Brown, University of Illinois) and a comparison of broadleaved forests (Orie Loucks, The Miami University [Ohio]). Data sets are available. Contact: Orie L. Loucks, The Miami University (513) 529-1677; Chinese PI: Li Wenhua, Commission on Integrated Survey of Natural Resources, CAS. |

SUMMARIES—*continued*

| Program | Lead U.S. Agency | Lead Chinese Agency | Dates | Objectives |
|---|---|---|---|---|
| South China Sea Paleoceanography | SCU | CAS | 1989- | Study of cores taken from the South China Sea to study paleoceanography and paleoclimate. Data available. Chinese PI: Luo Youlang, CAS South China Sea Institute of Oceanography, Guangzhou (86-20) 445-1335 x825; U.S. PI: Douglas Williams, Department of Geological Science, University of South Carolina (803) 777-7525. |
| South China Seas Sediment Analysis | NIU | CAS | | South China Sea Institute of Oceanology is conducting O-isotope work on sediment cores from the South China Sea. Chinese PI: Luo Youlang, SCSIO, (86-20) 445-1335 x825; U.S. PI: Hsin Yiling, Northern Illinois University, Department of Geology, (815) 703-7951. |

## MULTILATERAL AND NON-U.S. BILATERAL RESEARCH PROGRAMS

| Program | Countries | Lead Chinese Agency | Objectives |
|---|---|---|---|
| Aerosol Particle Analysis | Japan, Korea | BNU | A network of stations is being set up in China, Korea, and Japan for aerosol sampling and analysis at Beijing Normal University and Keio and Kyoto Universities in Japan. Contact: Zhu Guanghua, Beijing Normal University. Japanese Contact: Yoshikazu Hashimoto, Keio University. |

| Name | Agency | Foreign Partners | Description |
|---|---|---|---|
| Australian Monsoon Experiment (AMEX) | SMA | Australia, United States | A subprogram of TOGA consisting of a 6-week experiment in 1987 aimed describing the broad-scale structure of the Australian monsoon system. Chinese PI: Chen Longxun, SMA, 86-1-831-2277 x2758); Foreign PI: Greg Holland, Bureau of Meteorology Research Centre, Melbourne, Australia, 613-669-4501. |
| Chemistry of Glacial Ice Cores | CAS | Japan, former USSR, France, Denmark, Switzerland, United States | Lanzhou Institute of Glaciology and Geocryology hosts international expeditions exploring the chemistry of ice cores from mid-latitude glaciers. Chinese PI: Yao Tandong, Lanzhou Institute of Glaciology and Geocryology, 86-9-312-6725 x328; Foreign PI: Lonnie Thompson, Byrd Polar Research Center, Ohio State University (614) 292-6652; Claude Lorius, Laboratoire de Glaciologie et Geophysique de l'Environnement, 33-7-642-5872 x144); H. Oeschger, Physikalisches Institut der Universität Bern, Bern, Switzerland, 46-316-5811. |
| China-Japan Chinese Cooperative Study on the Kuroshio Ocean Current | SOA | Japan | This project is carried out at the Second Institute of Oceanography, Hangzhou. Chinese PI: Yuan Yaocu, SIO, 86-571-87-6924 x352; Foreign PI: Japanese Science and Technology Agency, Office of Ocean Technology and Development. |
| China-Japan Friendship Environmental Protection Center | NEPA | Japan | Agreement between the premiers of China and Japan for the establishment of a national data and information center and a national monitoring standards center at the National Environmetal Protection Agency. Japanese participation is coordinated by the Japanese Ministry of Foreign Affairs and the Environment Agency of Japan. Chinese Contact: Chen Zijiu, NEPA, 86-1-601-2118; Foreign Contact: Kazu Kato, Environment Agency, 81-33580-4982. |
| Cooperative Ecological Research Program | CAS | Germany | Collaboration under the UNESCO MAB Program, with eight different research projects in different parts of China. Chinese PI: Zhao Xianying, Chinese National Committee for MAB, 86-1-329-7418. Foreign PI: Dr. B. von Droste, CERP Coordinator, UNESCO, Paris, France, 33-1-45-68-40-67. |

## MULTILATERAL AND NON-U.S. BILATERAL PROGRAMS—continued

| Program | Countries | Lead Chinese Agency | Objectives |
|---|---|---|---|
| East Asian/North Pacific Regional Study (APARE) | Australia, Hong Kong, Japan, Korea, Taiwan, United States | SMA | East Asia/North Pacific Regional Study of the IGAC Program. Oversees PEM-West studies of atmospheric cycles of carbon, nitrogen, ozone, sulfur, and aerosols over the Pacific Basin. |
| Effects of Acid Rain on Forests and Lakes | Japan | RCEES | Research Center for Eco-Environmental Sciences acid rain study in Chongqing and other sites in southern China. Chinese PI: Feng Zongwei, RCEES, 86-1-256-1870; Foreign PI: Norio Ogura, Tokyo University of Agriculture and Technology, Department of Environmental Science and Resources, 81-423-34-6906. |
| Equatorial Mesoscale Experiment (EMEX) | Australia, United States | SMA | This project is carried out in conjunction with AMEX, a TOGA subprogram consisting of six week experiment in 1987 aimed at obtaining a description of the mesoscale convection components of the Australian monsoon system. Chinese PI: Chen Longxun, SMA, 86-1-831-2277 x2785; Foreign PI: Peter Webster, Penn State University, Department of Meteorology (814) 865-6840. |
| Sino-Japanese Atmosphere–Land Surface Processes Experiment | Japan | CAS | Known locally as the HEIFE experiment, it is a large study of land surface climatology and hydrology in the Heihe River Basin in Gansu Province. The objectives of the program are to investigate air-surface exchanges, energy and water budget, the boundary layer structure, distribution of atmospheric dust over the desert, and water requirements of crops. The project is a contribution to the Global Energy and Water Cycle Experiment and Biological Aspects of the Hydrological Cycle research programs. Chinese PI: Gao Youxi, Lanzhou Institute of Plateau Atmospheric Physics, 86-931-25311; Foreign PI: Y. Mitsuta, Kyoto University (81-774) 32-3111. |

| | | | |
|---|---|---|---|
| Rice Paddy Methane Studies | Germany | CAS | Cooperation between the Fraunhofer Institute for Atmospheric Environmental Research and the Institute of Atmospheric Physics consisting of continuous flux measurements carried out on rice fields in Hangzhou since 1987, using an automatic sampling and analyzing system. Data sets available. Chinese PI: Wang Mingxing, Institute of Atmospheric Physics, (86-1) 491-9851; Foreign PI: Dr. Seiler, Fraunhofer Institute. |
| Study of Atmospheric and Environmental Change at the South Pole | United States, Uruguay | CAS | Study of the atmospheric and environmental change at the South Pole from the late Pleistocene Epoch to the present. Chinese PI: Liu Tungsheng, Xi'an Laboratory of Loess and Quaternary Geology (86-29) 51773, Qin Dahe, Lanzhou Institute of Glaciology and Geocryology. |

# D

# Ecological Stations of the Chinese Academy of Sciences

CAS' long-term plan is to include all of its stations in CERN. However, initial implementation in the next 5 years will be limited to 29 stations which are identified below by highlighted text. "Leading" stations are further identified by an asterisk. The establishment of CERN has prompted a comprehensive reassessment of its stations and, consequently, this listing is subject to revision as CERN plans progress.

### STATIONS

Ailaoshan Forest Ecosystem Experiment Station, Kunming Institute of Ecology

*Ansai Comprehensive Soil and Water Conservation Experiment Station** (temperate semimoist, semiarid; Loess Plateau), Northwest Institute of Soil and Water Conservation

Bayanbulak Grassland Experiment Station, Xinjiang Institute of Biology, Pedology, and Desert Research

Beijing Agroecology Experiment Station, Institute of Geography

**Beijing Forest Ecosystem Experiment Station** (warm temperate deciduous broadleaved forest; (Mentougou District, Beijing), Institute of Botany

Cele Comprehensive Desert Management Experiment Station, Xinjiang Institute of Biology, Pedology, and Desert Research

*Changbaishan Forest Ecosystem Experiment Station (temperate broadleaved, mixed coniferous forest; northeastern China), Shenyang Institute of Applied Ecology

Changshu Agroecology Experiment Station (central subtropical hilly red earth zone), Nanjing Institute of Soil Science

Changwu Comprehensive Soil and Water Conservation Experiment Station (temperate semiarid; Loess Plateau, Shaanxi Province), Northwest Institute of Soil and Water Conservation

*Dayawan Marine Biology Experiment Station (subtropical marine bay; Guangzhou), South China Sea Institute of Oceanology

Dinghushan Subtropical Forest Ecosystem Experiment Station (subtropical evergreen broadleaved forest; Guangzhou), South China Institute of Botany

*Donghu Lake Ecosystem Experiment Station (subtropical freshwater lake; Wuhan), Wuhan Institute of Hydrobiology

Donshan Aquaculture Experiment Station, Nanjing Institute of Geography and Limnology

*Fengqiu Comprehensive Agroecology Experiment Station (warm temperate, semimoist, semiarid; Huang–Huai–Hai Plain, Henan Province), Nanjing Institute of Soil Science

*Fukang Desert Ecosystem Experiment Station (temperate desert) Xinjiang Institute of Biology, Pedology, and Desert Research

Gonggashan Alpine Ecosystem Experiment Station (subalpine coniferous forest), Chengdu Institute of Mountain Hazards and Environment

Guizhou Karst Mountain Ecosystem Experiment Station, Guiyang Institute of Geochemistry

Guyuan Ecological Experiment Station, Northwest Institute of Soil and Water Conservation

*Haibei Alpine Meadow Ecosystem Experiment Station (temperate alpine meadow; Qinghai Province), Northwest Plateau Institute of Biology

Hailun Agroecology Experiment Station (temperate, semimoist), Heilongjiang Institute of Agricultural Modernization

Hainan Tropical Marine Biology Experiment Station, South China Sea Institute of Oceanology

*Heshan Comprehensive Downland Experiment Station (subtropical evergreen broadleaved forest; Guangdong Province), South China Institute of Botany

Huangdao Mariculture Experiment Station (temperate marine bay; Northeast China), Qingdao Institute of Oceanology

Huitong Subtropical Forest Ecosystem Experiment Station (subtropical evergreen broadleaved forest), Shenyang Institute of Applied Ecology

\*Inner Mongolia Grassland Experiment Station (temperate grassland; near Xilinhot), Institute of Botany

Jiulianshan Forest Ecology Experiment Station, Commission for Integrated Survey of Natural Resources

Linze Desert Experiment Station, Lanzhou Institute of Plateau Atmospheric Physics

**Luancheng Comprehensive Agroecology Experiment Station** (warm temperate, semimoist, Huang–Huai–Hai Plain), Shijiazhuang Institute of Agricultural Modernization

Maowen Forest Ecosystem Experiment Station, Chengdu Institute of Biology

Maowusu Ecology Experiment Station, Institute of Botany

Mosuowan Desert Experiment Station, Xinjiang Institute of Biology, Pedology, and Desert Research

**Naiman Comprehensive Desert Management Experiment Station** (warm temperate, semiarid; Keerqing Desert), Lanzhou Institute of Desert Research

Nanpi Agroecology Experiment Station (Huang–Huai–Hai Plain), Shijiazhuang Institute of Agricultural Modernization

**Qianyanzhou Comprehensive Red Soil Hill Experiment Station** (central subtropical, hilly red earth), Commission for Integrated Survey of Natural Resources

**Sanjiang Plain Marshland Experiment Station** (temperate three-river plain and wetlands; Northeast China), Changchun Institute of Geography

Shantou Marine Botany Experiment Station, South China Sea Institute of Oceanology

**Shapotou Desert Experiment Station** (temperate semidesert; near Zhongwei, Inner Mongolia Autonomous Region), Lanzhou Institute of Desert Research

\***Shenyang Ecological Experiment Station** (warm temperate, semimoist; Liao River Plain), Shenyang Institute of Applied Ecology

Taihangshan Mountain Ecology Experiment Station, Shijiazhuang Institute of Agricultural Modernization

Taihu Agroecology Experiment Station (subtropical freshwater lake), Nanjing Institute of Soil Science

**Taihu Lake Comprehensive (Agroecology) Experiment Station** (subtropical freshwater lake), Nanjing Geography and Limnology Institute

Talimu Water Balance Experiment Station, Xinjiang Institute of Geography

**Taoyuan Agroecology Experiment Station** (central subtropical hilly red earth), Changsha Institute of Agricultural Modernization

Turpan Hongqi Desert Experiment Station, Xinjiang Institute of Biology, Pedology, and Desert Research
Wulanaodu Grassland Experiment Station (temperate; Inner Mongolia Autonomous Region), Shenyang Institute of Applied Ecology
Wuxihongchiba Artificial Subtropical Alpine Grassland Experiment Station, Commission for Integrated Survey of Natural Resources
**Xiaoliang Artificial Tropical Forest Ecosystem Experiment Station**, (tropical rainforest), Kunming Institute of Ecology
**Xishuangbanna Tropical Forest Experimental Station** (tropical rain forest; Yunnan Province), Kunming Institute of Ecology
Xinming Forest Ecology Experiment Station, Shenyang Institute of Applied Ecology
**Yanting Purple Soil Agroecology Experiment Station** (central subtropical hilly purple soil; Sichuan Province), Chengdu Institute of Mountain Hazards and Environment
*****Yingtan Red Soil Hill Experiment Station** (northern subtropical, moist), Nanjing Institute of Soil Science
**Yucheng Comprehensive Experiment Station** (warm temperate, semimoist, semiarid; Huang–Huai–Hai Plain), Institute of Geography
Zhanjiang Marine Animal Experiment Station, South China Sea Institute of Oceanology

## NETWORK STUDY ON ECOSYSTEMS IN CHINA[2]

### Study on Structure and Function of Main Ecosystems in China and Approaches of Increasing Their Productivity (1991–1995)[3]

1. **General Objectives**
   1) Establish long-term observational and research networks on
      a. observation and monitoring of environmental changes and biological aspects.
      b. important ecological processes of ecosystems and impacts of human activities on them.
      c. data quality control and information system.
   2) Study structure, function, and dynamics of main ecosystems; test and develop theories of ecology.
   3) Investigate approaches to improving ecosystem management.
2. **Tasks**
   1) Network agroecology study
      a. Create a GIS for research areas.
      b. Establish station-level managerial model of representative

ecosystem(s) and study water condition, nutrient cycling, energy flow, and management.
   c. Study energy-saving agricultural inputs and study their long-term impacts on environment.
   d. Study some important ecological processes, for example, decomposition and accumulation of organic matters in soil, degradation and accumulation of pollutants, methane ($CH_4$) fluxes from soil and its regulation.

All of these research activities will be carried out at Hailun, Shenyang, Fengqiu, Yingtan, Ansai, Fukang ["leading" agroforestry/agroecosystem CERN stations] and other CERN stations.

2) Network forest ecosystem study
   a. Create geographic information systems (GIS) for research areas.
   b. Study structure, function, and dynamics of forest ecosystems, including population structure and dynamics, energy exchange, nutrient cycling, and hydrological and meteorological functions.
   c. Make ecological models describing structure, function, and dynamics of ecosystems.
   d. Establish managerial models of ecosystems.

All of these research activities will be carried out at Changbaishan, Beijing, Huitong, Dinghushan, and Heshan [all are "leading" forestry CERN stations except Huitong]

3) Network study on grassland ecosystems
   a. Create GIS for research areas.
   b. Study structure, productivity, food web, and energy flow of ecosystems.
   c. Study carbon, nitrogen, phosphorus, and sulfur cycling and water balance.
   d. Establish managerial model of ecosystems.

All of these research activities will be carried out at the Inner Mongolia and Haibei stations.

4) Study on lake and coastal ecosystems
   a. Create GIS for research areas.
   b. Study the food web and energy flow and their relationship to productivity of ecosystems.
   c. Study carbon, nitrogen, and phosphorus cycling between different energy levels and surface of ecosystem and their effects on environment and productivity.
   d. Impacts of human activity on structure and function of ecosystem and their regulation.
   e. Establish managerial model of ecosystems.

5) Study of ecosystem study techniques and information systems
   a. Standardize observational methods and publish related handbooks. Study methodology of data quality control and operation of technical systems at station and research centers.
   b. Data management. Make software suitable for operating information system at research stations. Make GIS and apply it in research and management of natural resources. Make models suitable for the research of structure, function, and dynamics of ecosystems and their regulation.

## NOTES

1. The overall listing was updated in April 1992. An update of "leading" station designations was received in September 1992.
2. Original translated by Zhao Shidong, Shenyang Institute of Applied Ecology.
3. Outline of the general objectives and tasks in the CERN research plan.

# E

# Contact Information for Selected Institutions

Beijing Forestry University
Xiaozhuang, Qinghua Dong Lu
Haidian District
100083 Beijing, China
Telephone: 86-1-256.8811
Cable: 9131 Beijing

Beijing Normal University
Institute of Low Energy Physics
100875 Beijing, China
Telephone: 86-1-201.1954
Fax: 86-1-201.3929

China Association for Science and
    Technology
86 Xue Yuan Nan Lu
100081 Beijing, China
Telephone: 86-1-831.8877
Fax: 86-1-832.1914

China Remote Sensing Satellite
    Ground Station
45 Beisanhuanxi Road
P.O. Box 2434
10086 Beijing, China
Telephone: 86-1-256.1214
Telex: 210222 RSGS CN

Chinese Academy of Meteorological
    Sciences
46 Baishiqiao Road
Western Suburb
10081 Beijing, China
Telephone: 86-1-832.7390
Fax: 86-1-832.7390, 861-831.1191
Telex: 22094 FDSMA CN
Cable: 2894

Chinese National Climate Committee
State Meteorological Administration
Department of Science and
    Technology Development
46 Baishiqiao Road
100081 Beijing, China
Telephone: 86-1-831.2277 x2632

Chinese National Committee for the IGBP
Chinese Academy of Sciences
Bureau of Resources and Environmental Sciences
52 Sanlihe Road
100864 Beijing, China
Telephone: 86-1-329.7534
Fax: 86-1-801.1095
Telex: 22474 ASCHI CN
Cable: 2233

Chinese National Committee for the Man and the Biosphere Program
Chinese Academy of Sciences
52 Sanlihe Road
100864 Beijing, China
Telephone: 86-1-329.7418
Fax: 86-1-801.1095
Telex: 22474 ASCHI CN
Cable: 2233

Commission for Integrated Survey of Natural Resources
Chinese Academy of Sciences
P.O. Box 767
100101 Beijing, China
Telephone: 86-1-446.551/438
Cable: 4844 Beijing

Guangzhou Institute of Geography
100 Xianlie Road
510070 Guangzhou, China
Telephone: 86-20-765.600/962
Cable: 1146

Guiyang Institute of Geochemistry
Chinese Academy of Sciences
73 Guanshui Lu
550002 Guiyang, China
Telephone: 86-851-25502
Cable: 7181 Guiyang

Institute of Atmospheric Physics
Chinese Academy of Sciences
P.O. Box 2718
10080 Beijing, China
Telephone: 86-1-256.2458
Fax: 86-1-256.2347

Institute of Botany
Chinese Academy of Sciences
141 Xizhimenwai Street
100044 Beijing, China
Telephone: 86-1-831.2840
Fax: 86-1-831.9534

Institute of Forestry Science
Chinese Academy of Forestry Science
Wanshoushan Hou
100091 Beijing, China
Telephone: 86-1-258.2211
Fax: 86-1-258.2317

Institute of Geography
Chinese Academy of Sciences
Building 917, Datun Road
Andingmenwai
100101 Beijing, China
Telephone: 86-1-491.4841
Fax: 86-1-256.6099
Cable: 9135

Institute of Hydrological Engineering
Ministry of Geology and Minerals
Zhengding Xian
050800 Shijiazhuang, Hebei, China
Telephone: 86-311-901.129

Institute of Karst Geology
Ministry of Geology and Minerals
40 Qixin Lu
541004 Guilin, Guangxi, China
Telephone: 86-773-44.2442
Fax: 86-773-44.3708

APPENDIX E

Institute of Zoology
Chinese Academy of Sciences
19 Zhongguancun Lu, Haidian
  District
100080 Beijing, China
Telephone: 86-1-255.1267
Fax: 86-1-256.5689

Lanzhou Institute of Glaciology and
  Geocryology
Chinese Academy of Sciences
174 Donggang West Road
730000 Lanzhou, China
Telephone: 86-931-48.5241
Fax: 86-931-48.5241

Lanzhou Institute of Plateau
  Atmospheric Physics
18 West Donggang Road
730000 Lanzhou, China
Telephone: 86-931-25311
Fax: 86-931-418.667

Ministry of Agriculture
Agricultural Society of China
11 Nongzhanguan Nanli
100026 Beijing, China
Telephone: 86-1-500.3366 x2248
Fax: 86-l-500.2448

Ministry of Foreign Affairs
225 Chaoyangmennei Dajie
100701 Beijing, China
Telephone: 86-1-553.831

Ministry of Geology and Mineral
  Resources
64 Funei Dajie
100812 Beijing, China
Telephone: 86-1-6O3.1144
Fax: 86-1-601.7791
Telex: 22531 MGMRC CN
Cable: 0966 Beijing

Ministry of Water Resources
Department of Water Resources
1 Baiguang Lu Ertiao
100761 Beijing, China
Telephone: 86-1-327.3322 x4721
Fax: 86-1-326.0365
Telex: 22466 CHMEP CN
Cable: 7193 Beijing

Nanjing Institute of Environmental
  Science
National Environmental Protection
  Agency
120 Jiangwangmiao Street
210042 Nanjing, China
Telephone: 86-25-761.374/3045
Fax: 86-25-406.104
Telex: 34025 ISSAS CN
Cable: 3883 Nanjing

Nanjing Institute of Geography and
  Limnology
Chinese Academy of Sciences
73 East Beijing Road
210008 Nanjing, China
Telephone: 86-25-711.864,
  86-25-631.441, 86-25-631.510
Fax: 86-25-714.759
Telex: 34025 ISSAS CN
Cable: 1472 Nanjing

Nanjing Institute of Soil Science
Chinese Academy of Sciences
71 East Beijing Road
P.O. Box 821
210008 Nanjing, China
Telephone: 86-25-303.562
  86-25-713.781
Fax: 86-25-712.663
Telex: 34025 ISSAS CN
Cable: 1099 Nanjing

Nanjing University
Department of Geo and Ocean
　Sciences
Nanjing, China
Telephone: 634651, ext. 2264
Fax: 86-25-306.387
Telex: 34151 PRCNU CN

National Climate Change
　Coordination Group
State Meteorological Administration
Division of Climate Resources
46 Baishiqiao Road
100081 Beijing, China
Telephone: 86-1-832.7340

National Environmental Protection
　Agency
115 Xizhimennei Nanxiaojie
Beijing, China
Telephone: 86-1-601.1186
Fax: 86-1-601.5641

National Natural Science
　Foundation of China
43 Baojia Street, Xicheng District
Beijing, China
Telephone: 86-1-654.523
Fax: 86-1-201.0306
Telex: 222434 NSFC CN
Cable: 6199

Northwest Institute of Soil and
　Water Conservation
Chinese Academy of Sciences
712100 Yangling, Shaanxi, China
Telephone: 09297.2412
Telex: 3932

Peking University
100871 Beijing, China
Telephone: 86-1-256.1166
Fax: 86-1-256.4095

Peking University Center of
　Environmental Sciences
100871 Beijing, China
Telephone: 86-1-256.1166 x3875
Fax: 86-1-250.1295

Qingdao Institute of Oceanology
Chinese Academy of Sciences
7 Nanhai Road
Qingdao, China
Telephone: 86-971-279.344
Telex: 32222 ISS CN
Cable: 3152 Qingdao

Research Center for Eco-
　Environmental Sciences
Chinese Academy of Sciences
P.O. Box 934
100083 Beijing, China
Telephone: 86-1-285.308
Fax: 86-1-284.898

Shanghai Institute of Plant
　Physiology
Chinese Academy of Sciences
300 Fenglin Road
20032 Shanghai, China
Telephone: 86-21-437.2090
Fax: 86-21-433.2385
Telex: 33275
Cable: 3232

Shenyang Institute of Applied
　Ecology
Chinese Academy of Sciences
72 Wenhua Road
110015 Shenyang, China
Telephone: 86-24-383.401 231
Fax: 86-24-391.320
Telex: 80095 IMRAS CN

South China Institute of Botany
Chinese Academy of Sciences
510650 Guangzhou, China
Telephone: 86-20-705.626

APPENDIX E

South China Sea Institute of
  Oceanology
Chinese Academy of Sciences
164 West Xingang Road
501301 Guangzhou, China
Telephone: 86-20-447.336
Cable: 0380

State Education Commission
Department of Science and
  Technology
37 Damucang, Xidan
100816 Beijing, China
Telephone: 86-1-602.0784
Fax: 86-1-602.0784

State Environmental Protection
  Commission Secretariat
National Environmental Protection
  Agency
115 Xizhimennei Nanxiaojie
100035 Beijing, China
Telephone: 86-1-832.9911 x3501
Fax: 86-1-601.1194

State Meteorological Administration
46 Baishiqiao Lu, Xijiao
100081 Beijing, China
Telephone: 86-1-891.679

State Oceanographic Administration
Department of Science and
  Technology
1 Fuxingmenwai Avenue
100860 Beijing, China
Fax: 86-1-803.3515

State Planning Commission
Department of National Land and
  Regional Affairs
38 Yuetan Nanjie
100824 Beijing, China
Telephone: 86-1-809.1605
Fax: 86-1-851.2929

State Science and Technology
  Commission
Department of Science and
  Technology for Social
  Development
54 Sanlihe Road
100862 Beijing, China
Telephone: 86-1-801.1847, 2163
Fax: 86-1-801.2594
Telex: 22349 SSTCC CN
Cable: 2233

Xi'an Laboratory of Loess and
  Quaternary Geology
Chinese Academy of Sciences
P.O. Box 17
Xi'an, Shaanxi, China
Telephone: 86-29-51773
Fax: 86-29-752.566

Xinjiang Institute of Biology,
  Pedology, and Desert Research
Chinese Academy of Sciences
40 South of Beijing Road
830011 Urumqi, Xinjiang, China
Telephone: 86-991-33.5642
Fax: 86-991-33.5459

Zhongshan University
Department of Atmospheric
  Sciences
Guangzhou, China
Telephone: 86-20-446.300/960

# F

# Abbreviations and Acronyms

| | |
|---|---|
| AMEX | Australian Monsoon Experiment |
| APARE | East Asia/North Pacific Regional Study |
| ASEAN | Association of Southeast Asian Nations |
| AVHRR | Advanced Very High Resolution Radiometer |
| BAHC | Biospheric Aspects of the Hydrological Cycle |
| CAMS | Chinese Academy of Meteorological Sciences |
| CAS | Chinese Academy of Sciences |
| CAST | Chinese Association of Science and Technology |
| CERN | Chinese Ecological Research Network |
| CFC | Chlorofluorocarbon |
| CHAASE | China and America Air–Sea Experiments |
| $CH_4$ | Methane |
| CISNAR | Commission for Integrated Survey of Natural Resources |
| CNCC | Chinese National Climate Committee |
| CNCIGBP | Chinese National Committee for the International Geosphere–Biosphere Program |
| COARE | Coupled Ocean–Atmosphere Response Agreement |
| CODATA | Committee on Data for Science and Technology |
| $CO_2$ | Carbon dioxide |
| CRAES | Chinese Research Academy of Environmental Sciences |

| | |
|---|---|
| CSCPRC | Committee on Scholarly Communication with the People's Republic of China |
| DIS | Data and Information Systems |
| DOE | Department of Energy (U.S.) |
| EAWEP | East Asia and Western Pacific |
| ENSO | *El Niño* Southern Oscillation |
| EPA | Environmental Protection Agency (U.S.) |
| GAIM | Global Analysis, Interpretation, and Modeling |
| GCM | General Circulation Model |
| GCTE | Global Change and Terrestrial Ecosystems |
| GEWEX | Global Energy and Water Cycle Experiment |
| GIS | Geographic Information System |
| GMCC | Global Monitoring for Climate Change Laboratory |
| HEIFE | Sino–Japanese Atmosphere–Land Surface Processes Experiment |
| HD/GEC | Human Dimensions of Global Environmental Change |
| ICSU | International Council of Scientific Unions |
| IGAC | International Global Atmospheric Chemistry Program |
| IGBP | International Geosphere–Biosphere Program |
| IPCC | Intergovernmental Panel on Climate Change |
| ISSC | International Social Science Council |
| JGOFS | Joint Global Ocean Flux Study |
| LASG | Laboratory of Numerical Modelling for Atmospheric Sciences and Geophysical Fluid Dynamics |
| LOICZ | Land–Ocean Interactions in the Coastal Zone |
| LREIS | Laboratory of Resources and Environment Information Systems |
| LTER | Long-Term Ecological Research |
| MAB | Man and the Biosphere Program |
| MFECS | Margin Flux in the East China Sea |
| MOA | Ministry of Agriculture |
| MOF | Ministry of Forestry |
| MOFA | Ministry of Foreign Affairs |
| MOGM | Ministry of Geology and Mineral Resources |
| MOWR | Ministry of Water Resources |

# APPENDIX F

| | |
|---|---|
| NAS | National Academy of Sciences |
| NASA | National Aeronautics and Space Administration |
| NCAR | National Center for Atmospheric Research |
| NCCCG | National Climate Change Coordination Group |
| NDVI | Normalized Difference Vegetation Index |
| NEPA | National Environmental Protection Agency (China) |
| NOAA | National Oceanic and Atmospheric Administration |
| $NO_2$ | Nitrite |
| $NO_x$ | Nitrogen oxides |
| $N_2O$ | Nitrous oxide |
| NPP | Net Primary Productivity |
| NSF | National Science Foundation |
| NSFC | National Natural Science Foundation of China |
| | |
| $O_3$ | Ozone |
| | |
| PAGES | Past Global Changes |
| PC | Personal Computer |
| PEM | Pacific Exploratory Mission |
| PIXE | Proton-Induced X–Ray Emission |
| | |
| QA/QC | Quality Assurance and Quality Control |
| | |
| RCEES | Research Center for Eco–Environmental Sciences |
| RRC | Regional Research Center |
| RRN | Regional Research Network |
| RRS | Regional Research Site |
| | |
| SC-IGBP | Scientific Committee for the International Geosphere–Biosphere Program |
| SEDC | State Education Commission |
| SEPC | State Environmental Protection Commission |
| SMA | State Meteorological Administration |
| SOA | State Oceanographic Administration |
| $SO_2$ | Sulfur dioxide |
| SPC | State Planning Commission |
| SSTC | State Science and Technology Commission |
| START | System for Analysis, Research, and Training |
| | |
| TM | Thematic Mapper |
| TOGA | Tropical Ocean and Global Atmosphere Program |

| | |
|---|---|
| UNCED | United Nations Conference on Environment and Development |
| UNDP | United Nations Development Program |
| UNEP | United Nations Environment Program |
| UNESCO | United Nations Educational Scientific and Cultural Organization |
| | |
| WCRP | World Climate Research Program |
| WDC | World Data Center |
| WMO | World Meteorological Organization |
| WOCE | World Ocean Circulation Experiment |

Currency exchange rate used: $1=5.5 *yuan*.

# Index

## A

Acid rain. *See* Precipitation composition
*Acta Scientiae Circumstantiae (Huangjing Kexue Xuebao [Journal of Environmental Sciences])*, 92, 161
*Ad hoc* USA–PRC Committee for the Joint Study of Global Change, 21
Advanced very high resolution radiometer (AVHRR) data, 8, 49, 63, 81, 104, 137, 159, 171
*Advances in Atmospheric Sciences*, 143
Aerosols, 10, 30, 57, 58, 59, 93, 95–96, 136, 184, 192
An Zhisheng, 169, 176, 190
Anhui Institute of Optics and Fine Mechanics, 58, 96
Antarctica, 30, 96, 149–150, 195
Arimoto, Richard, 116, 169, 190
Asian Development Bank, 51
Astronomy data, 41
*Atlas of Climate Physics of Tropical Pacific Ocean*, 77
Atmospheric chemistry. *See also* International Global Atmospheric Chemistry (IGAC) Project
  overview of research in, 92–93
  studies in specific areas of, 9–11, 93–97, 117, 183, 187
  and U.S. Science and Technology Umbrella Agreement, 183–185
Atmospheric Chemistry Center, 138
Atmospheric deposition, 11, 96–97, 187, 194
Augustana College, 106
Australia, 76, 144, 193, 194

## B

Beijing Agroecology Experiment Station, 144, 197
Beijing Astronomical Observatory, 41
Beijing Data Center, 138
Beijing Forest Ecosystem Experiment Station, 113, 143, 197

Beijing Forestry University, 203
Beijing Normal University
  contact information for, 203
  initiative in global change education, 47–48
  Institute of Low-Energy Physics, 135–136, 192
Beijing University. *See* Peking University
Berry, Joseph, 19
Bilateral projects. *See also* Multilateral and non-U.S. bilateral research programs under U.S. Science and Technology Umbrella Agreement, 182–192
Biogeochemistry. *See also* Trace gases
  research in, 11–12, 91, 104, 106, 107, 162, 163, 169, 190
Biological and Geological Survey, U.S., 140
Biospheric Aspects of the Hydrological Cycle and Global Energy and Water Cycle Experiment (BAHC/GEWEX), 28, 35, 89, 177
  explanation of, 65
  institutes conducting research related to, 53–54
  research conducted on, 65–69
  Sino–Japanese Atmosphere–Land Processes Experiment (HEIFE), 35, 89, 108, 150, 194
Brown, Sandra, 88, 166, 191
Brunig, Hans, 88, 166
Brunt, James, 190
Bureau of Meteorology Research Centre, 193

## C

Cang Shuxi, 176
Cao Jiaxin, 157
Cao Jiping, 32
Cao Pifu, 32
Carbon dioxide–vegetation interactions. *See* Climate–vegetation interactions
Carnegie Institution, Stanford University, 166
Cen Jiafa, 32
Center for Environmental Sciences, 158
Center of Computation, 43
Chen Guodong, 147
Chen Guofan, 32
Chen Jiaqi, 26, 153
Chen Junchang, 178
Chen Kai, 159
Chen Lieting, 174
Chen Longxun, 184, 193, 194
Chen Panqin, 26, 32
Chen Qinglong, 26
Chen Quangong, 82
Chen Shixun, 167, 171, 175
Chen Shupeng, 26
Chen Tegu, 167
Chen Yongning, 26
Chen Yuejuan, 173
Chen Zijiu, 193
Cheng Guodong, 61
Cheng Jicheng, 159
Cheng Tianwen, 179
China and American Air–Sea Experiments (CHAASE), 96, 169
China Association for Science and Technology (CAST), 25, 38, 203
*China Environmental Science (Zhongguo Huanjing Kexue)*, 43, 92
China Ocean Press, 50, 70
China Remote Sensing Satellite Ground Station
  contact information for, 203
  discussion of, 136–137
Chinese Academy of Meteorological Sciences (CAMS), 35, 37, 38, 49
  contact information for, 203
  discussion of, 138–139
  research at, 93, 95, 98, 138–139, 183–185

INDEX

Chinese Academy of Sciences
    (CAS), 37, 38, 65, 75, 81, 112,
    187–189, 190, 191, 192, 193,
    194–195
  ecological stations of, 197–202
  function of, 5, 40–41
  funding of global change
    research by, 40
  initiative in global change
    education, 47–48
  reports and articles by, 82
  research priorities of, 3–4, 19, 28
Chinese Ecological Research
    Network (CERN), 5, 8–9, 28, 53,
    113, 115, 140, 145, 151, 190
  cooperation with U.S. Long-Term
    Ecological Research Network,
    87–88
  discussion of, 83–89
  field stations of, 85, 197–200
  information system of, 85–86
  research of, 86–89, 111, 112, 200–202
  subcenters of, 84–85
  Synthesis Center of, 84–86, 113, 115, 140
Chinese National Climate
    Committee (CNCC), 18, 76
  contact information for, 204
  establishment of, 4, 32
  organization and membership of, 32–34
  research programs of, 34–35
Chinese National Committee for
    the International Geosphere–
    Biosphere Program
    (CNCIGBP), 5, 8, 18
  assistance to panel from, 19
  *Bulletin of the CNCIGBP*, 29, 60
  contact information for, 204
  function of, 3–4
  organization and membership of, 25–28
  and projects relevant to IGBP, 30–32
  research agenda of, 28–32, 111–112

  workshops of, 29–30
Chinese Research Academy of
    Environmental Sciences
    (CRAES), 35, 37, 38, 43, 93, 137, 186
Chinese Universities Remote
    Sensing Center, 159. *See also*
    Institute of Remote Sensing
    Technology and Application
Chlorofluorocarbons (CFCs), 16,
    42–43, 185, 186
Chou Jifan, 32
*Climate*, 34
Climate–biosphere interactions, 112
Climate change. *See* also National
    Climate Change Coordination
    Group; Past Global Changes
  effecting land cover change, 12,
    108–116
  impact on economic
    development, 24, 80
  responses to, 36, 51
Climate change impacts, 4, 35, 42,
    36, 139, 190–191
Climate Institute, 190
Climate models, 35, 68, 78, 138, 188
*Climate Monitoring Bulletin*, 76
*Climate Physics in the Tropical Ocean,
    Atlas of*, 77
Climate Research Center, 138
Climate research program. *See*
    Chinese National Climate
    Committee
Climate–vegetation interactions, 39,
    62–65, 104–105, 164, 191
Climatology, land surface and
    hydrology, 65–69, 89
Coal
  dependence on, 15, 16
  emissions from burning, 9, 121,
    122, 187
Coal-bed methane, 42, 187
Coastal zone. *See also* Land–Ocean
    Interactions in the Coastal
    Zone (LOICZ); Marine
    environments research
    in China, 69–70

definition of, 72
Collaboration
  benefits of, 13
  panel views on, 119, 121–124
  with Australia, 76, 144, 193, 194
  with Denmark, 193
  with former Soviet Union, 148, 149, 158, 193
  with France, 193
  with Germany, 87, 162, 166, 193, 195
  with Hong Kong, 194
  with Japan, 43, 116, 136, 137, 148, 162, 192–194. *See also* Sino-Japanese Atmosphere-Land Surface Processes Experiment (HEIFE)
  with Korea, 192, 194
  with Switzerland, 193, 195
  with Taiwan, 194
  with United Kingdom, 144
  with United States, 75, 76, 87–88, 94, 96, 136, 142, 144, 148, 153, 158, 162, 166, 193–195
  with Uruguay, 150, 195
Colorado State University, 87, 184
Columbia-Lamont Doherty Geological Observatory, 185
Commission for Integrated Survey of Natural Resources (CISNAR)
  contact information for, 204
  discussion of, 139–141
  research conducted at, 41, 83, 84, 113
Commission on Atmospheric Chemistry and Global Pollution, 56
Committee on Data for Science and Technology (CODATA), 81
Committee on Scholarly Communication with the People's Republic of China (CSCPRC), 16–18, 88, 156
Conservation technologies, 42–43
Cooperative Ecological Research Project, 88, 166, 193
Coupled Ocean–Atmosphere Response Agreement (COARE), 75, 76
Cui Haiting, 26, 157, 158, 176
Cui Zhijiu, 157

# D

Data. *See also* World Data Center (WDC-D); AVHRR
  access to observational stratosphere ozone, 96
  access to precipitation, 97
  agricultural, 140
  Chinese Ecological Research Network, 85–86
  climate, 34, 143, 188
  climate research subprogram on, 34–35
  collection, management, and accessibility of Chinese, 39, 82, 120–121, 122
  Commission for Integrated Survey of Natural Resources, 139–140
  data management training, 190
  forestry, 171
  linkage between DIS and regional and national centers responsible for, 82
  National Environmental Protection Agency, 43
  oceanographic, 189
Data and Information Systems (DIS) for the IGBP, 8, 28, 145
  description of, 81–82
  institutes conducting research related to, 54
Deng Genyun, 173–174
Deng Nan, 26, 32, 51
Denmark, 193
Department of Agriculture, U.S., 182–183
Department of Energy, (U.S.)–Chinese Academy of Sciences (DOE–CAS) Joint Research on the Greenhouse Effect, 57, 89, 102, 142, 146, 187–189

Department of Science and
Technology for Social
Development. *See* State Science
and Technology Commission
Ding Yang, 164
Ding Yihui, 27, 138, 184, 185
Dinghushan Subtropical Forest
Ecosystem Experiment Station,
165, 166, 198
Donahue, Thomas, 48
Dong Yuanhua, 152, 155
Dozier, Jeff, 149
Dust, 10, 15, 67, 95
study of long-range transport of,
5, 58, 116, 169, 190

**E**

East Asia/North Pacific Regional
Study (APARE), 58–59, 94, 194
East China Normal University, 47
Eddy, John, 17
Education, 47–48, 80, 122
Ehleringer, James, 166
Eighth 5-Year Plan (1991–1995), 4,
29–30, 51
*El Niño* monitoring, 76–77
*El Niño* Southern Oscillation, 142,
148, 149, 168, 188
Ellis, James, 111
Emissions, 92–95. *See also* Coal
biogenic, 2–3, 15, 103–104
data collection regarding, 39, 122
greenhouse gas, 5, 42, 51, 57, 184,
*See also* DOE-CAS
industrial, 2–3
*Environmental Chemistry (Huangjing Huaxue)*, 92, 161
Environmental Protection Agency,
U.S., 42, 112, 152, 185–187
*Environmental Science (Huangjing Kexue)*, 92, 161
*Environmental Science and Technology Information (Huanjing Keji Qingbao)*, 43
*Environmental Science Research (Huanjing Kexue Yanjiu)*, 43

**F**

Fan Xinqin, 157
Fang Weiqing, 27, 32
Fang Zongyi, 32
Feng Sijian, 32
Feng Zongwei, 113, 194
Fengqiu Comprehensive
Agroecology Experiment
Station, 154, 198
Field, Chris, 166
First Institute of Oceanography, 50,
70, 159
Fossil energy. *See* Science and
Technology Umbrella
Agreement, U.S.; Coal
France, 193
Fraunhofer Institute, 57, 102, 195
Fu Baopu, 32
Fu Congbin, 27, 48, 81, 104
Fu Wei, 164
Fujimura, Mitsuru, 136
Fukang Desert Ecosystem
Observation and Experiment
Station, 170
Funding, 39–40, 43, 44, 45, 46, 119–
120, 122, 161

**G**

Galloway, James, 19, 21, 191
Gan Shijun, 51
Gan Zijun, 32, 168
Gansu Grassland Project, 81–82
Gansu Grasslands Ecological
Research Institute, 81–82, 110
Gao Youxi, 32, 194
General circulation models, 35, 39,
67, 142, 187, 190
Geographic information systems,
63, 85, 114, 115, 137, 145, 159
Geophysical data, 41
Georgia Institute of Technology, 94
Germany, 162, 166, 193, 195
Gifford, Roger, 164
Glaciers. *See* Ice core studies;
Lanzhou Institute of Glaciology

and Geocryology; World Data
    Center-D
Global Analysis, Interpretation, and
    Modeling (GAIM)
  description of, 78
  institutes conducting research
    related to, 54
  Special Committee, 28
Global change. *See also* Climate
    change
  China's impact on, 24, 97
  China's role in, 15–16
  China's view of, 2–3, 23–24, 65
  conceptual model of past, 29
  sensitive areas of environmental
    change and detection of early
    signals of significant, 29–30
Global Change and Terrestrial
    Ecosystems (GCTE) project, 28,
    108, 144, 191
  China's vegetation and climate
    and, 62–63
  explanation of, 62
  institutes conducting research
    related to, 54
  research related to, 29, 63–64
Global change education, 47–48, 80,
    122
Global change programs
  benefits of collaboration with
    Chinese in, 13
  Chinese participation in
    international, 5–12
  overview of, 24–25
  panel views regarding, 119–124
  workshops to implement China's,
    29–30
Global Energy and Water Cycle
    Experiment (GEWEX). *See*
    Biospheric Aspects of the
    Hydrological Cycle
Global Environment Facility (GEF),
    42, 79
Global Trends Network, 187
Gosz, James, 21, 88, 190
Grasslands, 31, 82, 86, 110–111. *See
    also* Land cover change

*Grasslands and Grassland Sciences in
    Northern China*, 111
Great Britain, 144
Greenhouse gases, 184. *See also*
    DOE-CAS; Emissions; IGAC;
    Trace gases
  grant to study control and
    mitigation of, 42, 51
Guangzhou Institute of Geography
  contact information for, 204
  discussion of, 141
  research conducted at, 73
Guiyang Institute of Geochemistry,
    149, 204
Guo Changming, 150
Guo Dexi, 32
Guo Shiqin, 27

**H**

Halons, 185, 186
Han Makang, 157
Han Shuti, 176
Hashimoto, Yoshikazu, 136
He Shanyu, 179
He Youhai, 168
Hebei Province General Hydrology
    Station, 101
Hehai University, 100, 101
Heshan Comprehensive Downland
    Experiment Station, 165, 198
Hing Ashi, 177
Historical analysis, 2, 3, 24, 32, 72–
    73, 109, 110, 113, 122, 138, 141,
    185, 188, 190. *See also* Past
    Global Changes
Hong Kong, 94, 194
Hong Yetang, 27
Hou Renzhi, 157
Hu Dunxin, 27, 32, 74, 160
Hu Huanling, 173
Hu Suozhong, 166
Huang Bingwei, 27
Huang Junxiong, 160
Huang Shengzhang, 179
Huang Wenhua, 158
Huang Zhenguo, 141, 177

INDEX 219

Huangdao Mariculture Experiment Station, 160, 198
Human Dimensions of Global Environmental Change Program (HD/GEC), 5, 79, 121
  Chinese participation in, 8
  discussion of, 53, 82
  impacts, 36
  institutes conducting research related to, 54
  research related to, 83, 179
Hydrology, 7, 11, 29, 98
  research conducted on, 64, 98–102, 164, 183

## I

Ice core studies, 100, 148–149, 193
Inayoshi, Akira, 136
Industrial emissions, 2–3
INFOTERRA, 161
Inner Mongolia Grassland Ecosystem Experiment Station, 89, 106, 143, 199
Institute of Analysis and Measurement, 43
Institute of Arid Regions Research, 138
Institute of Atmospheric Environment, 43
Institute of Atmospheric Physics, 35, 39, 57, 58, 85
  *Annual Report of the Institute of Atmospheric Physics*, 143
  contact information for, 204
  discussion of, 141–142
  research conducted at, 89, 93, 96, 98, 100, 142–143
  *Scientia Atmospherica Sinica (Daqi Kexue)*, 92, 143
Institute of Botany, 39
  contact information for, 204
  discussion of, 143
  ecological modeling at, 85
  research conducted at, 104, 114, 115, 144

terrestrial ecosystems studies, 6, 63–65
Institute of Desalination, 50
Institute of Ecological Environment, 43
Institute of Environmental and Economic Policy Research, 43
Institute of Environmental Information, 43
Institute of Environmental Law Research, 43
Institute of Environmental Management, 43
Institute of Environmental Standards, 43
Institute of Forestry Science, 204
Institute of Geography, 60, 85, 115
  contact information for, 204
  Department of Hydrology, 72
  discussion of, 144–145
  Laboratory of Resources and Environment Information Systems, 114–115, 144–145
  maps compiled by, 82
  research conducted at, 89, 98, 101, 114, 115, 145–146
Institute of Geophysics, 41
Institute of Hydrological Engineering, 204
Institute of Karst Geology, 204
Institute of Low-Energy Physics, 135–136
Institute of Marine Environmental Protection, 50, 70
Institute of Marine Scientific and Technological Information, 50, 70
Institute of Ocean Technology, 50, 70
Institute of Remote Sensing Applications, 137
Institute of Remote Sensing Technology and Application (Peking University), 137, 158–159
Institute of Systems Science, 83
Institute of Water Environment, 43

Institute of Zoology, 205
Integrated Research Center for Natural Resources and Agricultural Development, 84, 113, 140
Inter-American Institute for Global Change Research, 79
Intergovernmental Panel on Climate Change (IPCC), 36
International Council of Scientific Unions (ICSU), 3, 24, 26, 28, 53, 73, 79
International Geographical Union, 101
International Geosphere-Biosphere Program (IGBP), 3, 5, 17, 24, 43, 71, 81, 121
   projects relevant to, 30–32, 53–82, 11, 112, 173–178
International Global Atmospheric Chemistry (IGAC) Project, 28
   East Asia/North Pacific Regional Study (APARE), 58–59, 94, 194
   focus of, 5, 56–57
   institutes conducting research related to, 54
   research related to, 57–59, 94, 173–179
International Hydrology Project (UNESCO), 149
International Social Science Council, 5, 54, 82

## J

Japan
   collaboration with, 43, 94, 116, 136, 137, 148, 162, 192–194. *See also* Sino-Japanese Atmosphere–Land Surface Processes Experiment (HEIFE)
   measurement of aerosols over, 95
Japanese Environmental Protection Agency, 94, 193
Japanese Ministry of Foreign Affairs, 193
Japanese Science and Technology Agency, 193
Jiang Youxu, 32
Joint Global Ocean Flux Study (JGOFS), 7, 28, 160, 168
   Chinese committee activities of, 73–74
   institutes conducting research related to, 54–55
   National Program on Margin Flux in East China Sea, 74–75
   objectives of, 72, 73
   research related to, 74–75, 178

## K

Keio University, 136, 192
Ki Xingsheng, 173
Kiang, C. S., 19
Kong Xiangru, 61
Korea, 94, 192, 194
Korean Ocean Research and Development Institute, 58
Kotlyakov, M., 149
Kukla, George, 149
Kuroshio Ocean Current, 30, 50, 74, 75, 178, 193
Kyoto University, 65, 136, 192, 194

## L

*La Niña*, 77
Laboratoire de glaciologie et geophsique de l'environnement, 193
Laboratory for Frozen Soil Engineering, National Key, 147
Laboratory for Quantitative Vegetation Analysis, 63; *also* reported as Laboratory for Quantitative Ecology, 144
Laboratory of Material Cycling in the Pedosphere, National Key, 154
Laboratory of Numerical Modeling for Atmospheric Sciences and

Geophysical Fluid Dynamics (LASG), National Key, 143
Laboratory of Resources and Environment Information Systems (LREIS), National Key, 114–115, 144–145
Land cover change
  climate change effects on, 12, 108–116
  dynamics of, 115–116
  overview of, 108–109
  research conducted on, 109–115
Land–Ocean Interactions in the Coastal Zone (LOICZ), 7, 28, 71–73, 77, 154, 168
  discussion of, 71–72
  institutes conducting research related to, 54–55
  research related to, 72–73, 177–178
Landsat, 136, 137
Lanzhou Institute of Desert Research, 72, 110
Lanzhou Institute of Glaciology and Geocryology, 41, 60, 61, 143, 151, 170
  contact information for, 205
  discussion of, 147–148
  research conducted at, 100, 148–150, 193
Lanzhou Institute of Plateau Atmospheric Physics, 35, 65, 143, 147
  contact information for, 205
  discussion of, 150–151
  research conducted at, 151
Lawrence Livermore National Laboratory, 89
Leach, Beryl, 19
Li Bingyuan, 149
Li Fuxian, 50
Li Jijun, 61, 149
Li Peiji, 149, 174
Li Pingri, 141, 178
Li Tieying, 45
Li Wenhua, 27, 89, 179, 191
Li Yongqi, 27

Li Yushan, 156
Li Zechun, 32
Liaoning Normal University, 70
Life-supporting environment, 4, 30
Lin Guangsong, 155
Lin Hai, 27
Liu Chaohai, 149
Liu Chunzheng, 32, 100, 101
Liu Gengnian, 157
Liu Hongliang, 43
Liu Jibin, 36
Liu Jihan, 157
Liu Jingyi, 48, 163
Liu, Shaw, 19, 183
Liu Shida, 174
Liu Shu, 27
Liu Tungsheng, 27, 89, 168, 176, 195
Liu Xinren, 100
Liu Yubin, 33
Liu Yuyan, 176
Loess, 60, 168
  Plateau, 30, 80, 121, 156, 169
*Loess, Environment and Global Change*, 170
London University, 105
Long-Term Ecological Research (LTER), U.S., 87–88, 89, 90, 170
Loucks, Orie, 89, 166, 191
Lu Jingting, 27
Lu Jiuyuan, 33
Lu Shouben, 27
Lu Ting, 135
Lue Youlang, 167

**M**

Ma Ainai, 159
Ma Bin, 179
Ma Peimin, 173
Ma Yang, 27
Man and the Biosphere (MAB) Program, 88–89, 165 191, 193, 204
Mao Handing, 179
Maousu Ecology Experiment Station, 143, 199

Margin Flux in the East China Sea (MFECS), 74–75
Marine Environmental Forecasting Center, National, 50, 58, 70, 96
Marine environments research. *See also* Joint Global Ocean Flux Study (JGOFS); Land–Ocean Interactions in the Coastal Zone (LOICZ); Tropical Ocean and Global Atmosphere (TOGA) Program
  discussion of, 7, 69–70
  institutions involved in, 54–55, 70–71
  under U.S. Science and Technology Umbrella Agreement, 189
McCarthy, James, 17
Meteorological data, 49
Methane emissions, 15, 57, 103, 157, 189
  from coal-bed, 42, 187
  from rice paddies, 15, 57, 103, 182, 186, 195
Miami University, 191
Ministry of Agriculture (MOA), 49, 70, 82, 113, 140, 182, 205
Ministry of Chemical Industry, 42
Ministry of Energy, 49, 187
Ministry of Foreign Affairs (MOFA), 35, 205
Ministry of Forestry, 49
Ministry of Geology and Mineral Resources (MOGM), 71, 205
Ministry of Light Industry, 185
Ministry of Public Safety, 185
Ministry of Urban and Rural Construction, 49
Ministry of Water Resources (MOWR), 11
  contact information for, 205
  function of, 98
  research conducted at, 99–100, 183
Monsoon research, 80, 139, 143, 148, 149, 184. *See also* TOGA
  Australian, 194

climate, 62, 76
East Asian, 76, 114
Montreal Protocol on Substances that Deplete the Ozone, 1, 16, 42, 43
Mooney, Harold, 17
Multilateral and non-U.S. bilateral research programs, 192–195. *See also* Collaboration; *individual countries*

## N

Nanjing Institute of Environmental Science, 43
  contact information for, 205
  discussion of, 152
  research conducted at, 114, 186
Nanjing Institute of Geography and Limnology, 61, 149, 155
  contact information for, 205
  discussion of, 152–154
  research conducted at, 72, 100, 113
Nanjing Institute of Hydrology and Water Resources, 98, 100, 101
Nanjing Institute of Soil Science
  contact information for, 205
  research conducted at, 85, 104, 106, 152, 154
Nanjing University, 70
  contact information for, 206
  Department of Geo and Ocean Sciences, 70, 73, 113, 155–156
  research conducted at, 93, 115
National Academy of Sciences, 16
National Aeronautics and Space Administration, U.S. (NASA), 58, 94, 182, 183
National Center for Atmospheric Research (NCAR), 89, 182, 189, 191
  Community Climate Model, 139
National Center for Global Change Research, 80
National Climate Change

Coordination Group (NCCCG),
  11, 36, 49, 51, 102, 111
  contact information for, 206
  organization of, 35–36
National Environmental Protection
  Agency (NEPA), 37, 38
  contact information for, 206
  function of, 5, 35, 42–43, 97, 116
  research conducted at, 93, 96–97,
    112, 152, 185–187, 191, 193
  Sino–Japan Friendship
    Environmental Protection
    Center, 43, 193
National key laboratory system, 52
National Meteorological Center, 34,
  35, 49
National Natural Science
  Foundation of China (NSFC), 6,
  38, 73, 90
  contact information for, 206
  function of, 5, 40, 43–45
  funding of global change
    research by, 40, 59, 112, 120,
    122, 173–179
  *Guide to Programs*, 44
  organization of, 44–45
National Oceanographic and
  Atmospheric Administration,
  U.S. (NOAA), 58, 76, 89, 94,
  159, 183, 187, 189, 191
National Remote Sensing Center,
  52, 159
National Satellite Meteorological
  Center, 49, 63
National Science Foundation (NSF),
  U.S., 2, 17, 18, 76, 87, 90, 184,
  185
Natural resources data, 41
Net primary productivity, 104–105
New Mexico State University, 182
Nie Baofu, 167, 175
Ningbo Oceanography School, 50,
  70
Nippon Environmental Pollution
  Control Center, 136
Nitrous oxide emissions, 103–104
North Korea, 155
Northern Illinois University, 192
Northwest Institute of Soil and
  Water Conservation
  contact information for, 206
  discussion of, 156–157
  research conducted at, 106

## O

Oak Ridge National Laboratory, 89
Ocean flux studies, 74–75. *See also*
  Joint Global Ocean Flux Study
  (JGOFS); Marine environments
  research
Ohio State University, 193
Open laboratory system, 41, 52
Oregon Graduate Institute of
  Science and Technology, 89, 189
Oregon State University, 87
Ouyang Ziyuan, 27
Ozone, 5, 10, 32, 56, 58, 96

## P

Pacific Exploratory Mission (PEM),
  58
  East Asia/North Pacific Regional
    Study (APARE), 58–59, 94, 194
  PEM-West, 58–59, 94, 95, 183
Pacific Science Association, 124
Paleo-environments. *See also* Past
  Global Changes
  studies in, 146, 157, 190, 192
  uniqueness of Chinese records of,
    6, 60, 61, 121
Pan Liangbao, 153
Pan Yushen, 61
Panel on Global Climate Change
  Sciences in China
  formation of, 18
  methodology of, 18–20
  research areas focused on by, 91
  views on collaboration prospects,
    123–124
  views on contributions to
    international research
    programs, 121–122

views on global change programs, 119–121
Past Global Changes (PAGES), 4, 28, 29, 59–60, 112, 153, 157, 167
  institutes conducting research related to, 55
  research related to, 60–62, 80, 109, 174–177, 190, 192
Pei Chunli, 135
Peking University, 48, 58
  Center for Environmental Sciences, 158, 206
  contact information for, 206
  Institute of Remote Sensing Technology and Application, 137, 158–159
  research conducted at, 93, 96, 157–158
Penn State University, 194
Physikalisches Institut der Universität Bern, 193
Porter, Stephen, 169, 190
Precipitation composition, 11, 31, 56, 92, 93, 96–97, 157, 158, 163, 191
Pu Peimin, 153

## Q

Qian Junlong, 153, 174
Qian Zhengxu, 178
Qin Dahe, 149–150, 195
Qin Guoxiong, 164
Qin Yunshan, 159
Qin Zenghao, 178
Qingdao Institute of Oceanology, 70, 72, 167
  contact information for, 206
  discussion of, 159–160
Qingdao Ocean University, 70, 159
Qinghai-Tibet Plateau, 51, 61, 62, 80, 113, 117, 122, 144, 146, 153, 183, 184
Qiu Guangwen, 33
Qu Geping, 42

## R

Reardon-Anderson, James, 111
Reiners, William, 19
Ren Mei-e, 141, 155–156, 167, 178
Research Center for Eco-Environmental Sciences (RCEES)
  contact information for, 206
  discussion of, 160–161
  research conducted at, 93, 97, 113, 161–163, 194
Rice paddy emissions. *See* Methane
Russia. *See* Soviet Union, former
Russian Academy of Sciences, 48, 158

## S

Sacklette, Dr., 167
Schimel, David, 19
Science and Technology Umbrella Agreement, U.S.
  agriculture protocol, 182–183
  atmosphere protocol, 183–185
  environment protocol, 185–187
  fossil energy protocol, 187–189
  marine/fishery protocol, 189
  summaries of selected university-level bilateral projects, 190–192
*Scientia Atmospherica Sinica (Daqi Kexue)*, 92, 143
Scientific institutions
  contact information for, 203–207
  funding of, 39–40
  organization of, 4–5, 13, 37–39, 120
Second Institute of Oceanography, 50, 70
Shanghai Institute of Plant Physiology
  agricultural ecosystems studies by, 6, 64, 65
  contact information for, 206
  discussion of, 163–164
  research conducted at, 1–5, 101, 164

INDEX

Shao Kesheng, 157
Shen Guoquan, 33
Shen Qiuxing, 27
Shen Shanmin, 27
Shen Wenxiong, 33
Shenyang Institute of Applied
    Ecology, 206
Shi Nianhai, 175
Shi Shaohua, 153
Shi Yafeng, 61, 100, 148–149, 153, 178
Sino–Japanese Atmosphere–Land Surface Processes Experiment (HEIFE), 35, 65–67, 89, 108, 150, 194
Snow cover studies, 149
Soil nutrient studies, 106
Song Jian, 26, 51
Song Xueliang, 176
South China Botanical Garden, 165
South China Institute of Botany, 89
    contact information for, 206
    discussion of, 165–166
South China Institute of Environmental Protection, 43
South China Sea Institute of Oceanology, 71, 72, 141
    contact information for, 207
    discussion of, 167
    research conducted at, 72, 167–168
Soviet Union, former, 148, 149, 158, 193
Space science data, 41
State Education Commission (SEDC), 38
    contact information for, 207
    function of, 45, 47–48
State Environmental Protection Commission (SEPC), 35, 38, 42
    contact information for, 207
    function of, 5, 49
State Meteorological Administration (SMA), 38, 39, 59, 75, 76, 138. See also Chinese Academy of Meteorological Science (CAMS)
    contact information for, 207
    function of, 5, 49, 57, 98, 116
National Meteorological Center, 34, 35, 49
National Satellite Meteorological Center, 49, 63
research conducted at, 93, 96–97, 183–185, 193, 194
State Oceanographic Administration (SOA), 37, 38, 71, 75, 96, 189, 193
    contact information for, 207
    function of, 5, 35, 49–50, 58
State Planning Commission (SPC), 38, 139, 144
    contact information for, 207
    function of, 5, 50
State Science and Technology Commission (SSTC), 4, 35, 38, 61
    contact information for, 207
    function of, 51–52
    funding of global change research by, 40, 144
    role of, 5, 36
State University of New York, Albany, 89, 184, 187
State University of New York, Stony Brook, 89, 187, 188
Su Weihan, 163
Su Weimin, 176
Sulfur dioxide. See Coal
Sun Guchou, 174
Sun Honglie, 27, 61, 88, 139
Sun Xiangjun, 175
Sun Yuke, 167
Switzerland, 193
Synthetic Aperture Radar, 137
System for Analysis, Research, and Training (START)
    CNCIGBP proposal to Standing Committee of, 8, 79–81
    description of, 79
    institutes conducting research related to, 55–56
Sze, Nien Dak, 19

## T

Taihu Agroecology Experiment Station, 154, 199
Taihu Lake Comprehensive Experiment Station, 103, 199
Taiwan, 94, 194
Tang Keli, 156
Tang Maocang, 33, 61
Tang Peisung, 143
Tang Xiaoyan, 48, 157, 158
Tao Shiyan, 33
Terrestrial ecosystems, 6, 29
Thematic Mapper (TM) imagery, 136–137
Third Institute of Oceanography, 50, 71
Thompson, Lonnie, 148, 193
Tianshan Glacier Research Laboratory, 147
Tianshan Glaciology Research Station, 147, 170
Tibetan Plateau. *See* Qinghai-Tibet Plateau
Tibetan Plateau Institute, 138
Tieszen, Larry, 106
Tokyo University of Agriculture and Technology, 194
Tongji University, 47, 70
Trace gases. *See also* PEM-West
  biotic controls on, 11–12, 102–108
  impact of human activities on biological sources of, 29
  monitoring of, 57–58
  research conducted on, 9–10, 93–95, 162, 183, 190
*Translation of Environmental Science (Huanjing Kexue Yicong)*, 43
Tropical Ocean and Global Atmosphere (TOGA) Program, 50, 58, 75–76, 96, 138, 168, 184, 185, 189, 193
  *El Niño* monitoring by, 76–77
  Equatorial Mesoscale Experiment, 194
  explanation of, 4, 7, 35, 75

  institutes conducting research related to, 56
  monsoon climate research of, 76
  observation programs of, 76
Tu Mengzhao, 165
Tu Qingying, 152

## U

United Kingdom, 144
United Nations Conference on Environment and Development, 50
United Nations Development Program (UNDP), 42
United Nations Educational, Scientific, and Cultural Organization (UNESCO), 87, 149, 193
United Nations Environment Program (UNEP), 105, 155
United States. *See also* Science and Technology Umbrella Agreement, U.S.
  collaboration with, 75, 76, 87–88, 94, 96, 123, 136, 142, 144, 148, 153, 158, 162, 166, 193–195
University of Arizona, 185
University of Maryland, 190
University of Miami, 191
University of Michigan, 48
University of Minnesota, 153
University of New Mexico, 21
University of North Carolina, 158
University of Rhode Island, 58, 116, 169, 190
University of Science and Technology of China, 41, 47, 58, 93, 96
University of South Carolina, 192
University of Utah, 166
University of Virginia, 21, 87, 191
University of Washington, 190
University of Wisconsin, 185
Uruguay, 150, 195

# INDEX

## W

Waliguanshan Atmospheric Baseline Observatory, 94
Wan Guojiang, 175
Wang, Wei-Chyung, 19
Wang Baocan, 178
Wang Binke, 156
Wang Guiqin, 173
Wang Huadong, 48
Wang Juemou, 33
Wang Lixian, 27
Wang Mingxing, 136, 189
Wang Nailiang, 157
Wang Pinxian, 176
Wang Ren, 27
Wang Shaowu, 33, 157
Wang Suming, 153
Wang Tao, 110
Wang Tianduo, 64, 164
Wang, Wei-Chyung, 19, 188
Wang Wenxing, 43
Wang Xinfu, 135, 136
Wang Xinmin, 136
Wang Yangzu, 33
Wang Yuanzhong, 33
Water resources. *See also* Hydrology
Water-Saving Agriculture on the North China Plain study, 64, 108, 164
Water Science Institute of the Yellow River Basin Commission, 100
Wei Dingwen, 173
Wen Jianping, 51
Wen Qizhong, 153
Weng Duming, 33
Wilson, William, 158, 187
Winchester, John, 19
World Climate Research Program (WCRP), 5, 44, 121
  explanation of, 3, 24
  research projects of, 35, 53
World Data Center (WDC-D), 35, 36, 41, 52, 81, 84, 113, 140, 145, 147
  listing of WDC-D subcenters, 90
World Meteorological Organization (WMO), 3, 4, 24, 53, 138
World Ocean Circulation Experiment (WOCE), 50, 73, 168
Wu Baozhong, 33
Wu Yuguang, 135
Wuhan Hydraulic and Electrical Engineering University, 98, 100

## X

Xia Xuncheng, 170
Xi'an Laboratory of Loess and Quaternary Geology, 60, 136, 190, 195
  contact information for, 207
  discussion of, 168
  research conducted at, 169–170
Xiao Lingmu, 175
Xiao Xuchang, 27
Xiaoliang Artificial Tropical Forest Ecosystem Experiment Station, 165, 166
Xie Xiangfang, 179
Xie Zichu, 148, 149
Xinjiang Institute of Biology, Pedology, and Desert Research
  contact information for, 207
  discussion of, 170–171
  research conducted at, 106
Xinjiang Institute of Environmental Research, 43
Xu Daquan, 164
Xu Deying, 27
Xu Haipeng, 157
Xu Jialin, 48
Xu Songling, 174
Xu Xiru, 159

## Y

Yan Hongmo, 49–50
Yang Daqing, 149
Yang Linzhang, 152, 155
Yang Wenhe, 33
Yang Xiangheng, 155

Yang Xiongli, 164
Yao Tandong, 148, 175
Ye Duzheng, 2, 3, 19, 23, 26, 28–30, 33, 40, 51, 142, 177
Ye Zhizheng, 176
Ying Jonsheng, 176
Yingtan Red Soil Hill Experiment Station, 154, 200
Yu Shuwen, 164
Yu Xixian, 157
Yu Zhouwen, 33
Yuan Daoxian, 28
Yuan Yaocu, 193
Yuan Yeli, 178
Yucheng Integrated Experiment Station, 144, 200

## Z

Zeng Qingcun, 33, 142, 187
Zhang Chijun, 177
Zhang De'er, 175
Zhang Haifeng, 69
Zhang Jiacheng, 34
Zhang Jijia, 34
Zhang Jin, 177
Zhang Peiyuan, 146, 188
Zhang Qiaoling, 34
Zhang Qiaomin, 167, 168
Zhang Shen, 28
Zhang Shifa, 100
Zhang Xiaoye, 136, 169
Zhang Xinshi (a.k.a. Chang Hsin-shih or David Chang), 28, 61, 63, 104, 144, 143 146, 191
Zhang Xuebin, 153
Zhang Yue, 28
Zhang Zonghu, 28
Zhao Dianwu, 163
Zhao Huanting, 167
Zhao Jianping, 190
Zhao Qiguo, 28, 154
Zhao Songling, 176
Zhao Xianping, 193
Zhao Zhongci, 157
Zheng Benxing, 175
Zheng Changsu, 153
Zheng Du, 61, 144, 146
Zheng Qinglin, 184
Zhong Gongfu, 179
Zhongshan University, 167, 171, 207
Zhou Guangzhao, 80
Zhou Lisan, 83
Zhou Xiaoping, 185
Zhou Xiuji, 34, 138, 183
Zhou Yinlin, 175
Zhou Zejiang, 152
Zhu Guanghua, 135, 136
Zhu Yuanzhi, 167
Zhu Zhenda, 110
Zhuang Yahui, 113, 161
Zou Jiahua, 50
Zou Jingmeng, 32, 49
Zou Renlin, 175
Zou Tingru, 175
Zu Haipeng, 158
Zuo Dakang, 144